Homework Helpers

Math Grade 3

As a parent, you want your child to enjoy learning and to do well in school. The activities in the *Homework Helpers* series will help your child develop the skills and self-confidence that lead to success. Humorous illustrations and diverse formats make the activities interesting for your child.

HOW TO USE THIS BOOK

- Provide a quiet, comfortable place to work with your child.
- Plan a special time to work with your child. Create a warm, accepting atmosphere so your child will enjoy spending this time with you. Limit each session to one or two activities.
- Make sure your child understands the directions before beginning an activity.
- Check the answers with your child as soon as an activity has been completed. (Be sure to remove the answer pages from the center of the book before your child uses the book.)
- The activities in this book were selected from previously published Frank Schaffer materials.
- Topics covered in this book are addition and subtraction facts, place value to thousands and ten thousands, comparison of larger numbers, addition and subtraction of two-, three-, and four-digit numbers, regrouping, computing time, making change, using graphs, basic multiplication and division facts, recognition and comparison of fractional parts, geometry, and problem solving.
- For additional multiplication practice activities, please see *Homework Helpers—Multiplication Grade 3* (FS-8142)

ISBN #0-86734-108-4
FS-8141 Homework Helpers—Math Grade 3

23740 Hawthorne Blvd., Torrance, CA 90505

This Book Belongs To

Name ______________________________

Date ______________________________

Think and Write

Write the answers.

A.	6 + 6	9 + 2	6 + 4	7 + 7	2 + 7	3 + 8
B.	3 + 9	7 + 6	4 + 6	6 + 9	4 + 4	9 + 9
C.	9 + 5	3 + 5	8 + 2	8 + 5	9 + 3	6 + 3
D.	4 + 5	9 + 1	6 + 8	4 + 9	6 + 7	9 + 4
E.	8 + 6	6 + 5	7 + 8	9 + 6	8 + 3	7 + 4
F.	3 + 7	8 + 4	5 + 7	3 + 6	7 + 5	5 + 6
G.	9 + 7	4 + 7	1 + 9	8 + 7	2 + 9	5 + 8
H.	2 + 8	7 + 9	4 + 8	7 + 3		
I.	8 + 9	8 + 8	5 + 9	9 + 8		

Subtraction Maze

Work each problem. To find a path to the end of the pie-eating contest, begin at START. Color all the boxes whose answers are even numbers.

		13 - 6	16 - 6	18 - 9		
	11 - 6	12 - 5	14 - 8	17 - 9	10 - 7	
13 - 4	12 - 9	15 - 8	7 - 4	12 - 4	16 - 7	13 - 8
11 - 8	16 - 9	10 - 4	13 - 7	11 - 5	14 - 9	8 - 3
14 - 7	12 - 3	12 - 8	15 - 6	10 - 9	9 - 2	10 - 3
17 - 8	15 - 7	13 - 5	11 - 4	9 - 4	12 - 7	8 - 5
START 14 - 6	11 - 7	14 - 5	9 - 6	10 - 5	7 - 2	11 - 2

Dinosaurs Galore

Count the hundreds, tens, and ones. Write the numbers on the dinosaurs.

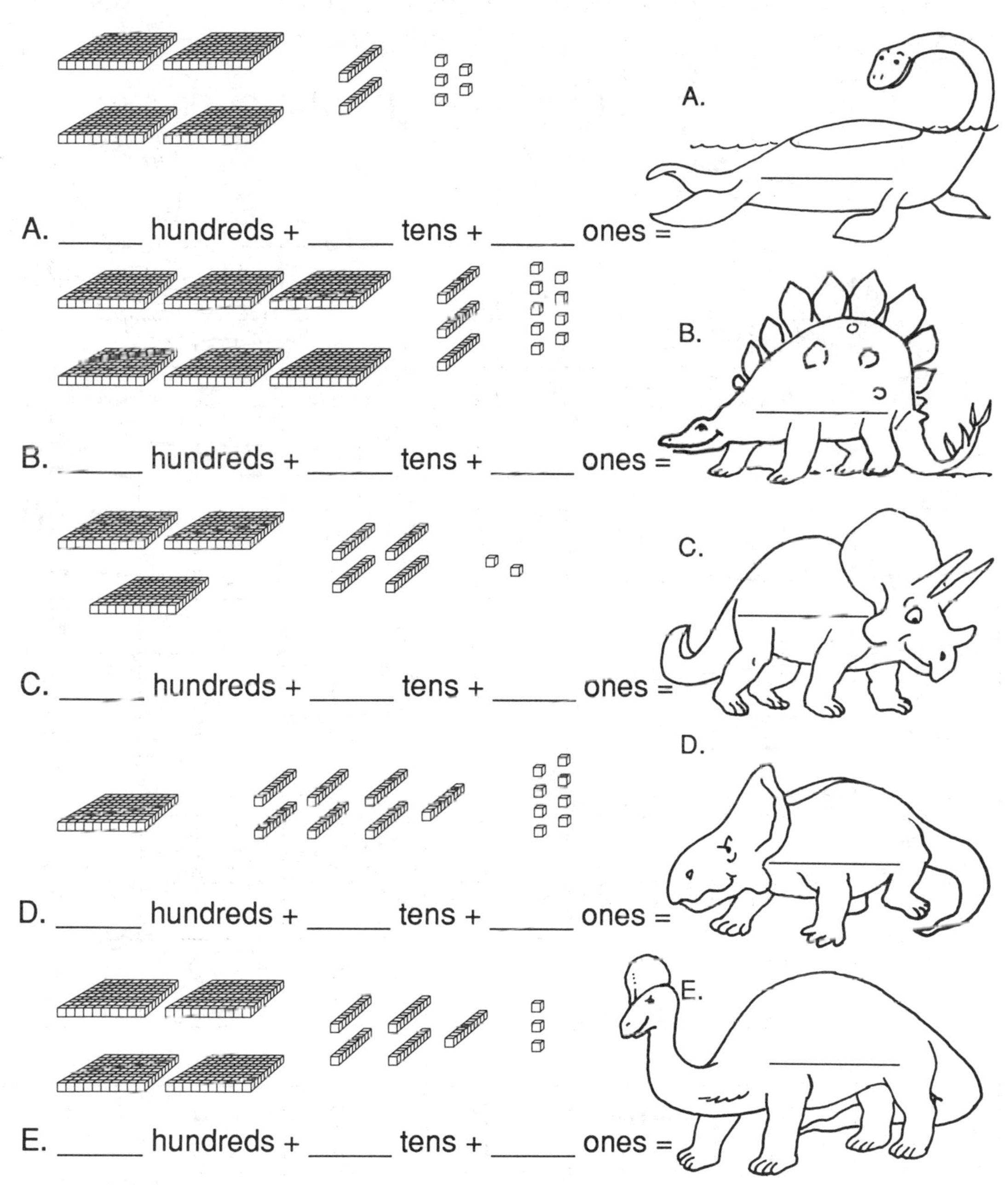

A. _____ hundreds + _____ tens + _____ ones =

B. _____ hundreds + _____ tens + _____ ones =

C. _____ hundreds + _____ tens + _____ ones =

D. _____ hundreds + _____ tens + _____ ones =

E. _____ hundreds + _____ tens + _____ ones =

Try This! Write the three-digit numbers used above in order from the largest to the smallest.

Beaver Delights

Fill in the blanks with **<** or **>**.
Color the logs with **>** brown.

< means *is less than* **>** means *is greater than*

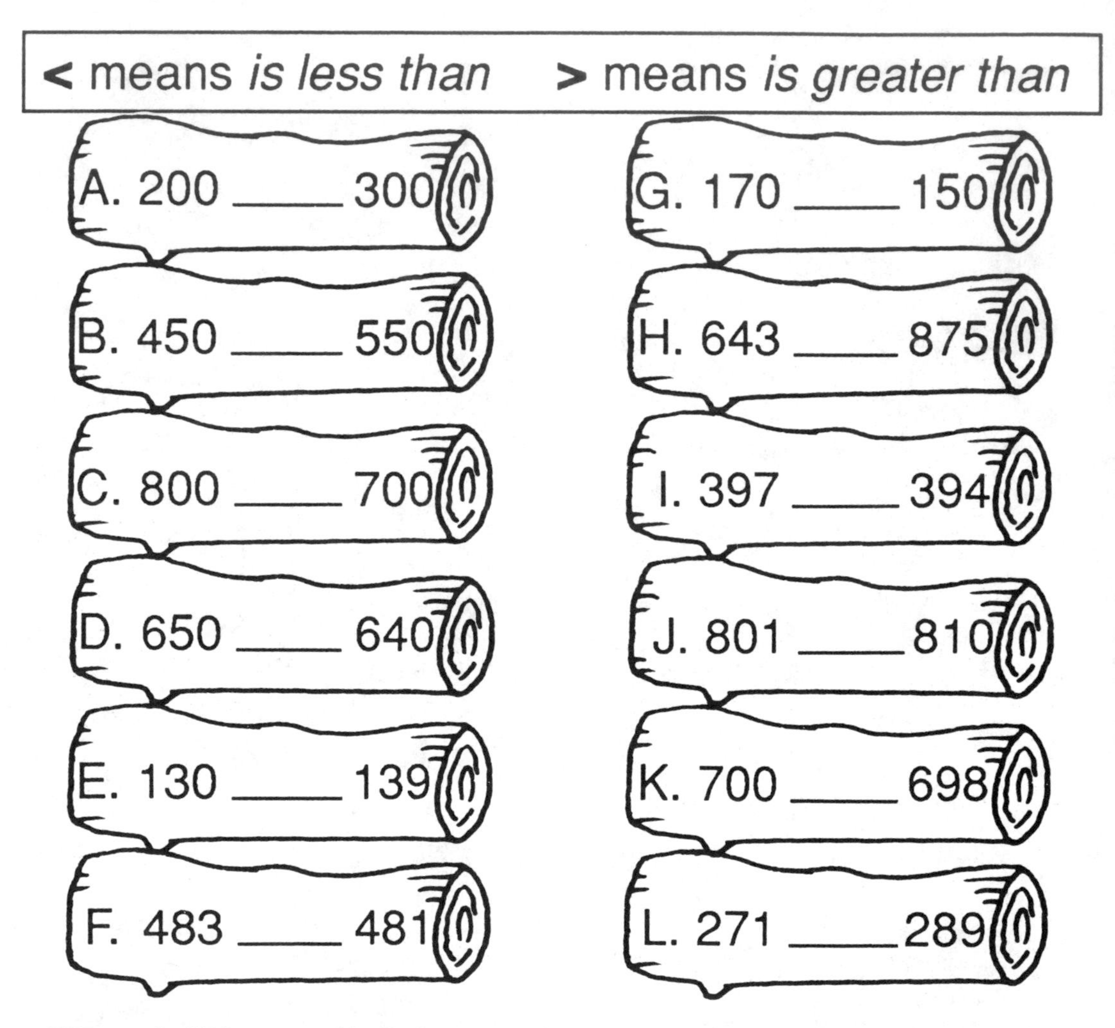

Try This! Write all 24 numbers above in order from the smallest to the largest.

Time to Expand

There were 3,245 visitors at the dinosaur museum last month.

The 3 represents **3,000** people.

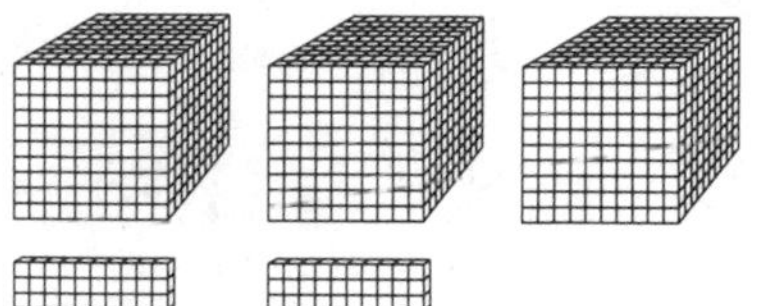

The 2 represents **200** people.

The 4 represents **40** people.

The 5 represents **5** people.

3,000 + 200 + 40 + 5 = 3,245

Circle the correct expanded form of each number.

A. 6,748
  a. 6,000 + 700 + 4 + 8
  b. 6,000 + 700 + 40 + 8
  c. 600 + 7,000 + 4 + 8
  d. 6,000 + 7 + 4 + 8

B. 8,153
  a. 8,000 + 100 + 50 + 3
  b. 8,000 + 10 + 50 + 3
  c. 800 + 100 + 5 + 3
  d. 800 + 100 + 50 + 3

C. 4,092
  a. 400 + 0 + 9 + 2
  b. 4,000 + 0 + 9 + 2
  c. 400 + 0 + 92
  d. 4,000 + 0 + 90 + 2

D. 2,960
  a. 2,000 + 90 + 6 + 0
  b. 200 + 990 + 60 + 0
  c. 2,000 + 900 + 60 + 0
  d. 200 + 900 + 60 + 0

Write the expanded form of the numbers below.

E. 9,589 ________ + ________ + ________ + ________

F. 5,266 ________ + ________ + ________ + ________

G. 8,631 ________ + ________ + ________ + ________

H. 3,075 ________ + ________ + ________ + ________

Try This! Find five four-digit numbers in your math book. Write them in expanded form and have a friend check them.

The Dinosaur Challenge

Take the Dinosaur Challenge. Write **>**, **<**, or **=** in each circle.

A. 12,468 ◯ 12,648

B. 59,428 ◯ 59,482

C. 91,030 ◯ 91,300

D. 82,512 ◯ 79,999

E. 43,999 ◯ 43,899

F. 38,060 ◯ 30,000 + 8,000 + 6

G. 51,010 ◯ 50,000 + 100 + 10

H. 10,770 ◯ 10,000 + 700 + 70

I. 73,111 ◯ 70,000 + 2,000 + 900

J. 96,002 ◯ 90,000 + 6,000 + 2

K. 40,000 + 9,000 + 900 + 9 ◯ 50,461

L. 10,000 + 9,000 + 70 + 7 ◯ 19,078

M. 30,000 + 3,000 + 600 + 3 ◯ 33,603

N. 80,000 + 1,000 + 500 ◯ 81,405

O. 60,000 + 2,000 + 800 + 60 ◯ 62,860

Write the letters of the problems that have **<** as their answer. ____ ____ ____ ____ ____

Unscramble the letters to make a color word. Color the dinosaur ____________________.

Try This! Write your zip code in expanded form.

Addition Garden Maze

Write the answers. To find the rabbit's path through the garden, begin at START. Color all the boxes with answers that are odd numbers.

START →	24 + 17	78 + 7	40 + 31	16 + 49	24 + 25
72 + 6	55 + 43	18 + 70	15 + 45	7 + 57	82 + 9
28 + 48	56 + 6	7 + 49	47 + 42	67 + 14	60 + 13
29 + 28	36 + 25	66 + 7	18 + 31	88 + 4	56 + 24
24 + 35	44 + 26	45 + 31	22 + 44	47 + 13	51 + 23
40 + 39	15 + 46	37 + 38	37 + 32	18 + 59	36 + 46
16 + 64	21 + 27	49 + 5	8 + 38	47 + 48	SEEDS

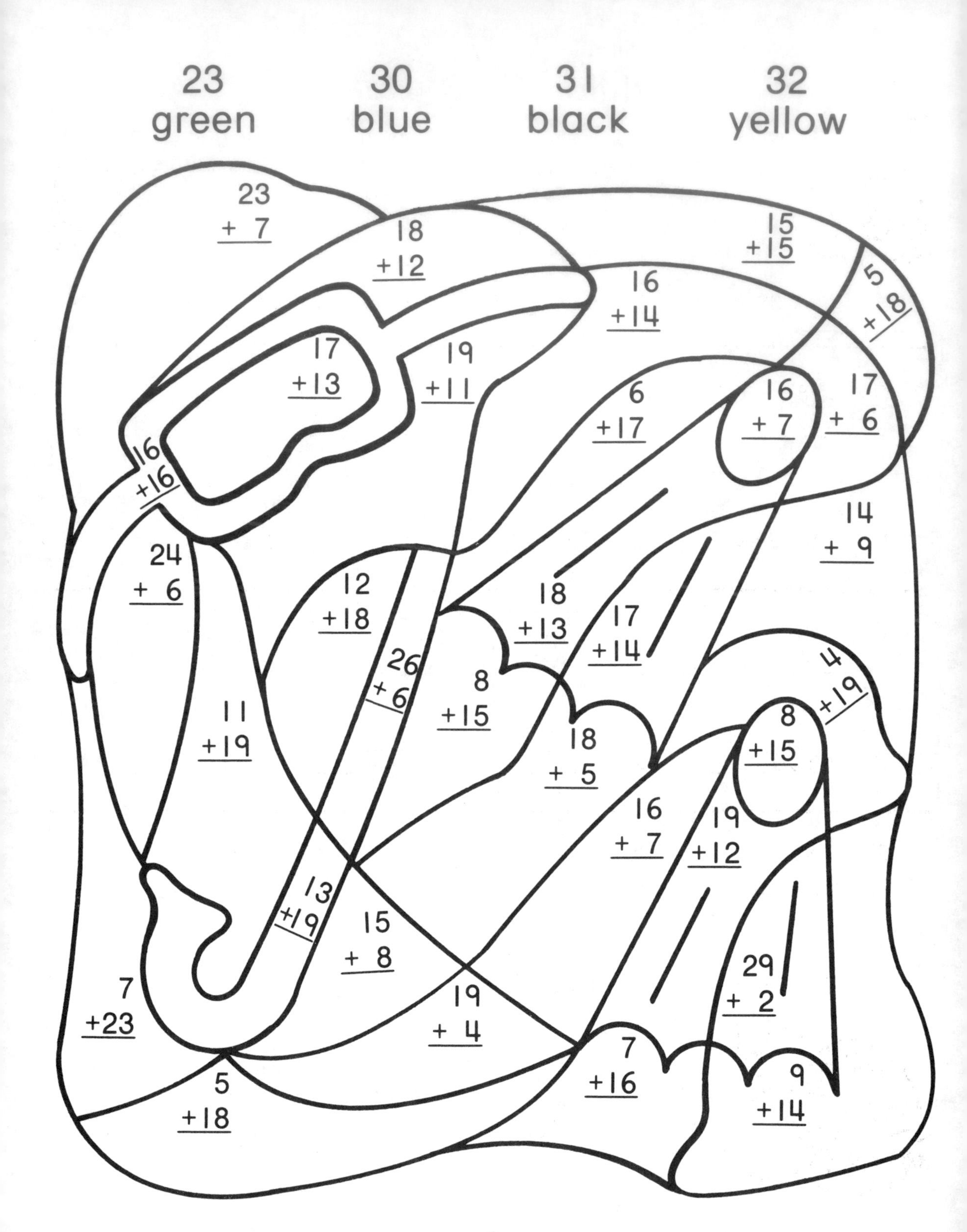
23
green
30
blue
31
black
32
yellow
23
+ 7
18
+12
15
+15
16
+14
5
+18
17
+13
19
+11
6
+17
16
+ 7
17
+ 6
16
+16
14
+ 9
24
+ 6
12
+18
18
+13
17
+14
26
+ 6
8
+15
4
+19
11
+19
18
+ 5
8
+15
16
+ 7
19
+12
13
+19
15
+ 8
29
+ 2
7
+23
19
+ 4
7
+16
9
+14
5
+18

Turtle Sums

Add.

A.

43	24	6	28
+ 17	+ 34	+ 67	+ 51

B. 25 + 33 = _____ 44 + 32 = _____ 80 + 17 = _____ 64 + 27 = _____

C.

38	17	51	47	83
+ 33	+ 42	+ 19	+ 17	+ 10

D. 18 + 37 = _____ 22 + 19 = _____ 31 + 36 = _____ 20 + 20 = _____

E.

28	28	55	49	24
+ 45	+ 68	+ 6	+ 49	+ 13

F. 29 + 26 = _____ 45 + 32 = _____ 36 + 38 = _____ 24 + 14 = _____

G.

47	18	33	45	61
+ 40	+ 39	+ 16	+ 17	+ 18

H. 28 + 49 = _____ 23 + 34 = _____ 39 + 4 = _____ 66 + 21 = _____

I.

25	75	78	59	49
+ 23	+ 15	+ 17	+ 37	+ 35

81 and **52** — **Yellow**
82 and **63** — **Brown**
83 and **71** — **Red**

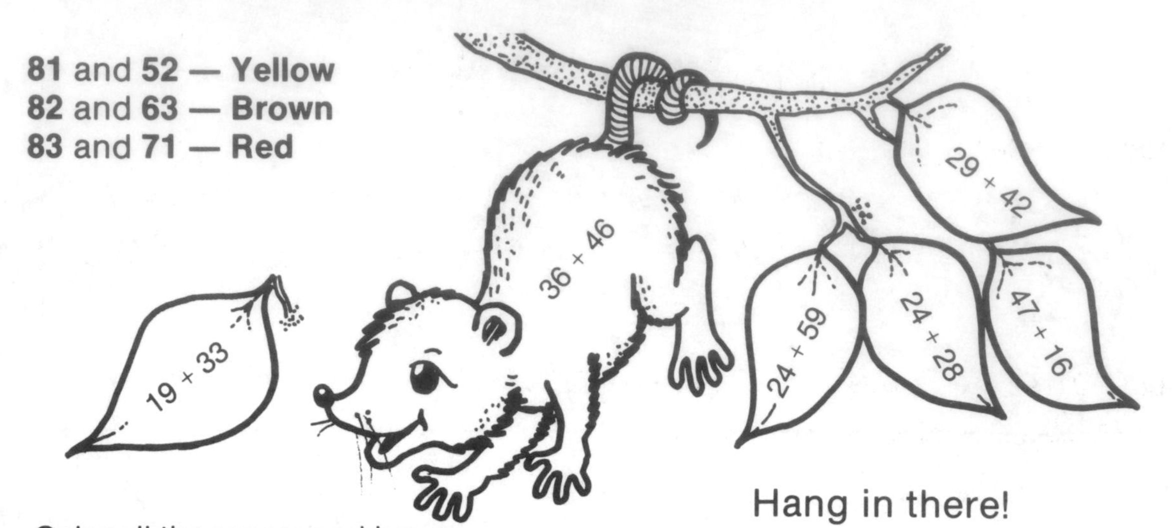

Color all the spaces and boxes.

17 + 66	59 + 12	48 + 23	49 + 32	38 + 44	64 + 17	29 + 42	58 + 23
54 + 17	23 + 29	56 + 25	44 + 37	39 + 13	43 + 28	28 + 55	45 + 38
26 + 57	43 + 28	59 + 24	22 + 59	35 + 28	36 + 16	33 + 38	68 + 13
16 + 65	38 + 14	55 + 28	54 + 27	13 + 69	33 + 48	46 + 25	34 + 47
38 + 45	65 + 18	29 + 42	52 + 29	49 + 14	14 + 67	42 + 29	24 + 28

Hanging by a Thread

Add and write the answers.

A.	384 + 532	291 + 210	464 + 65	327 + 492	193 + 555	186 + 761
B.	212 + 94	636 + 171	543 + 282	238 + 590	480 + 463	487 + 162
C.	429 + 80	330 + 384	688 + 51	346 + 393	480 + 284	487 + 122
D.	575 + 71	370 + 548	261 + 97	342 + 560	156 + 92	298 + 520
E.	175 + 683	874 + 94	647 + 270	253 + 253	486 + 141	254 + 80
F.	174 + 333	459 + 270	796 + 133	271 + 555	550 + 392	437 + 490
G.	542 + 266	358 + 461	460 + 460	242 + 492		

Brainwork! Use the digits *3*, *4*, *5*, *6*, *7*, and *8* to write a three-digit addition problem that needs regrouping. Then solve it.

Regrouping Race

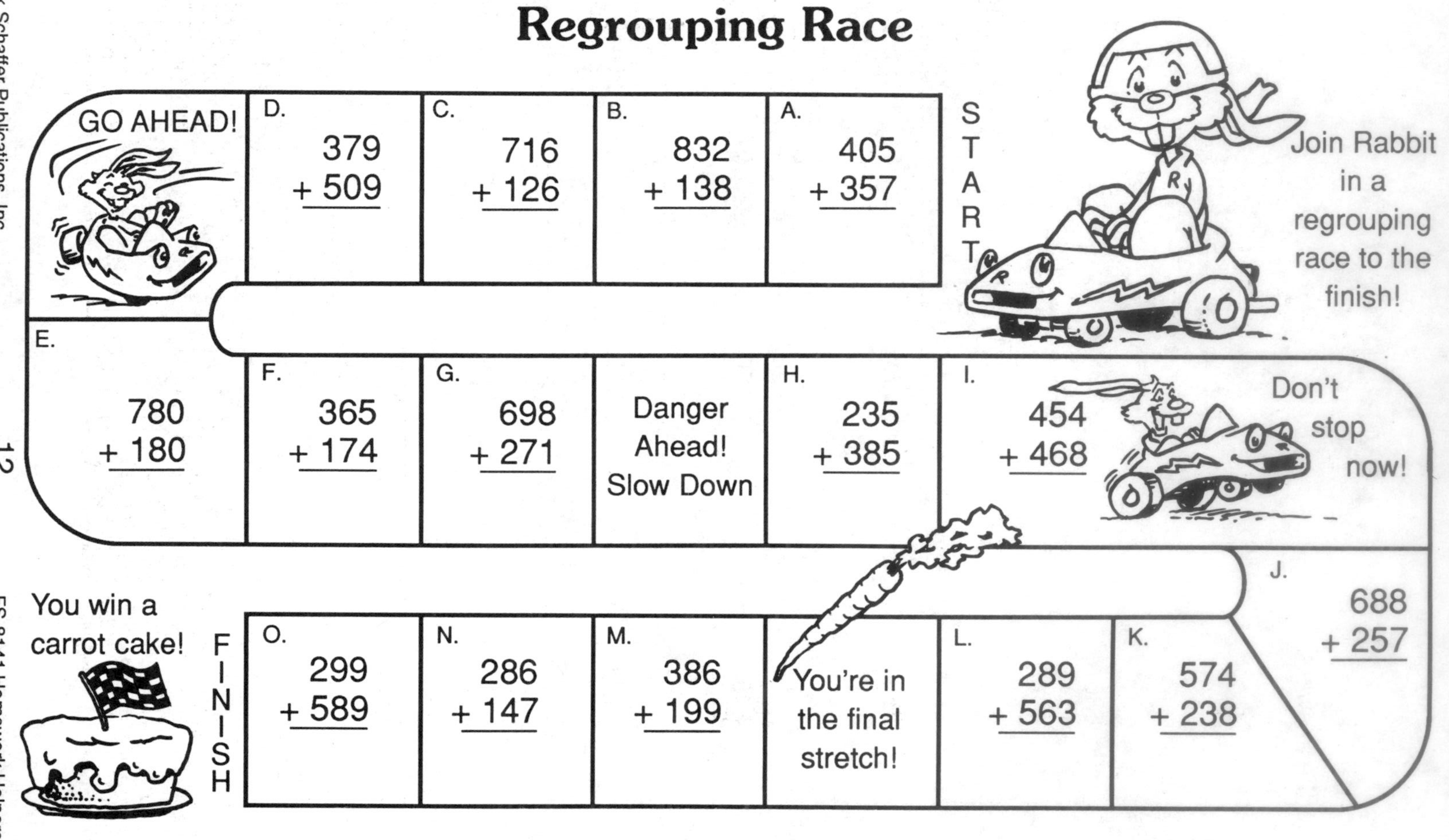

Try This! You went pretty fast around those curves! Prove that problem E is correct. Here's how. Begin with the answer. Now subtract from it one of the numbers you added. Your answer should be the other number!

Which Doghouse?

Write the answers. Then color the path to the doghouse where Spot will hide her bone. Color all the boxes with answers that are even numbers.

	START →	836 + 138	485 + 306	334 + 439	185 + 484
247 + 244	478 + 751	726 + 428	576 + 272	134 + 392	364 + 384
407 + 548	294 + 295	665 + 592	383 + 354	156 + 893	328 + 190
325 + 346	108 + 346	648 + 748	742 + 880	115 + 415	428 + 628
961 + 854	597 + 221	906 + 79	527 + 106	372 + 331	285 + 20
447 + 448	661 + 553	544 + 118	235 + 247	793 + 73	603 + 522
724 + 757	369 + 408	466 + 561	850 + 693	377 + 893	249 + 116

Diving for Answers

Write the answers.

A.	284 + 376	392 + 548	124 + 798	377 + 636	819 + 395	447 + 275	256 + 256
B.	443 + 488	155 + 655	792 + 488	684 + 679	185 + 336	758 + 84	675 + 75
C.	519 + 388	609 + 449	477 + 74	766 + 766	245 + 97	248 + 269	456 + 277
D.	504 + 96	777 + 567	199 + 299	143 + 157	924 + 76	761 + 189	498 + 197
E.	373 + 329	123 + 98	828 + 676	444 + 298	586 + 586	254 + 947	675 + 125

Brainwork! Circle two answers that have three digits. Add those two numbers together.

Chimp Challenge

Write the answers.

A.	74 213 + 112	324 210 + 707	316 550 + 22	760 432 + 24	49 72 + 116	601 126 + 50
B.	217 128 + 150	208 194 + 334	112 113 + 511	202 322 + 62	201 242 + 501	660 90 + 907
C.	522 127 + 227	26 448 + 121	51 153 + 45	666 121 + 401	140 220 + 101	55 346 + 123
D.	221 231 + 262	146 140 + 190	128 428 + 481	714 281 + 102	227 52 + 311	84 312 + 302
E.	22 622 + 132	837 262 + 231	242 285 + 51	454 14 + 404	102 102 + 141	346 12 + 491

Brainwork! Write three numbers whose sum is 1,000.

Adding Larger Numbers

A.	2,714 + 3,281	4,265 + 5,135	6,223 + 290	2,227 + 1,462	5,540 + 1,628
B.	4,175 + 377	1,097 + 1,983	7,142 + 2,355	8,426 + 722	1,014 + 792
C.	6,514 + 1,922	3,555 + 1,671	892 + 3,419	5,001 + 4,232	1,660 + 2,237
D.	1,554 + 667	8,817 + 1,012	5,572 + 918	1,090 + 703	2,813 + 2,763
E.	4,272 + 624	1,456 + 1,417	5,915 + 1,775	1,500 + 1,458	7,254 + 746
F.	2,754 + 2,787	2,861 + 668	7,565 + 1,907	4,841 + 1,827	5,817 + 2,062
G.	1,533 + 1,316	2,376 + 1,849	4,458 + 998	8,814 + 1,092	3,464 + 3,721

Early Risers

Subtract and write the answers.

A. 71 − 33	50 − 25	77 − 8	45 − 18	74 − 28	82 − 37
B. 90 − 47	91 − 19	93 − 38	74 − 57	30 − 21	71 − 39
C. 67 − 8	54 − 48	80 − 19	52 − 18	85 − 38	54 − 28
D. 70 − 26	78 − 59	43 − 7	75 − 66	54 − 15	93 − 9
E. 66 − 19	93 − 29	93 − 18	51 − 39	80 − 13	97 − 19

Brainwork! Write a subtraction word problem about a farm.

Slip-and-Slide Subtraction

Take a ride on the Slip-and-Slide. Solve each problem on your way.

Try This! Make a big splash by creating a subtraction slide with eight problems.

Prizewinning Pig

Find the path that leads to the prize Petunia won at the fair. First subtract and write the answers. Then color the boxes with odd answers.

→ START		78 − 57	94 − 9	55 − 17	58 − 10
73 − 19	39 − 11	67 − 17	58 − 33	97 − 19	62 − 6
81 − 29	45 − 9	98 − 22	88 − 20	70 − 26	36 − 18
42 − 17	58 − 41	62 − 59	96 − 45	74 − 16	82 − 20
60 − 15	87 − 11	57 − 9	81 − 17	48 − 10	61 − 7
74 − 13	36 − 29	28 − 13	43 − 35	74 − 26	98 − 46
84 − 26	47 − 31	57 − 24	99 − 27	44 − 18	72 − 18

Brainwork! Write at least three subtraction problems that have 45 as their answer.

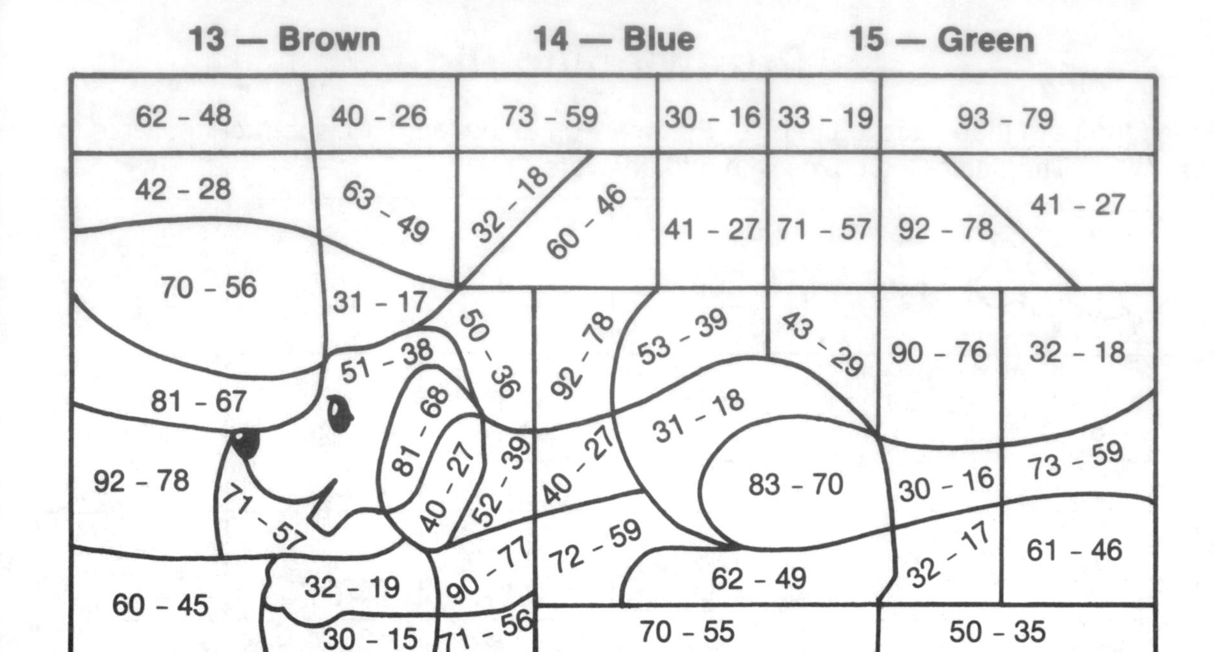

Color all the spaces and boxes.

93 - 79	72 - 59	61 - 48	81 - 67	63 - 49	62 - 49	91 - 77	53 - 39
60 - 46	51 - 37	82 - 69	91 - 78	50 - 37	70 - 57	41 - 27	80 - 66
43 - 29	71 - 57	31 - 18	61 - 47	52 - 38	81 - 68	91 - 77	50 - 36
30 - 15	92 - 77	74 - 59	62 - 47	54 - 39	60 - 45	72 - 57	83 - 68

Regrouping Challenge

Solve each problem. Then use the key to decode Rabbit's message.

122	214	239	253	268	277	384	388	446	595	619	623
O	R	I	Y	B	S	U	F	D	P	T	A

651 – 437	990 – 367	595 – 327	792 – 524	564 – 325	873 – 254
214					
R					

829 – 590	568 – 291

758 – 163	393 – 179	518 – 396	658 – 274	708 – 262

720 – 598	673 – 285

841 – 588	611 – 489	643 – 259

Try This! If A=1, B=2, and so on, what does *BED* – *ACE* equal? Make up a "word" problem like this. Use letters *A* through *I* (1 through 9) only.

Crowing for Corn

Write the answers.

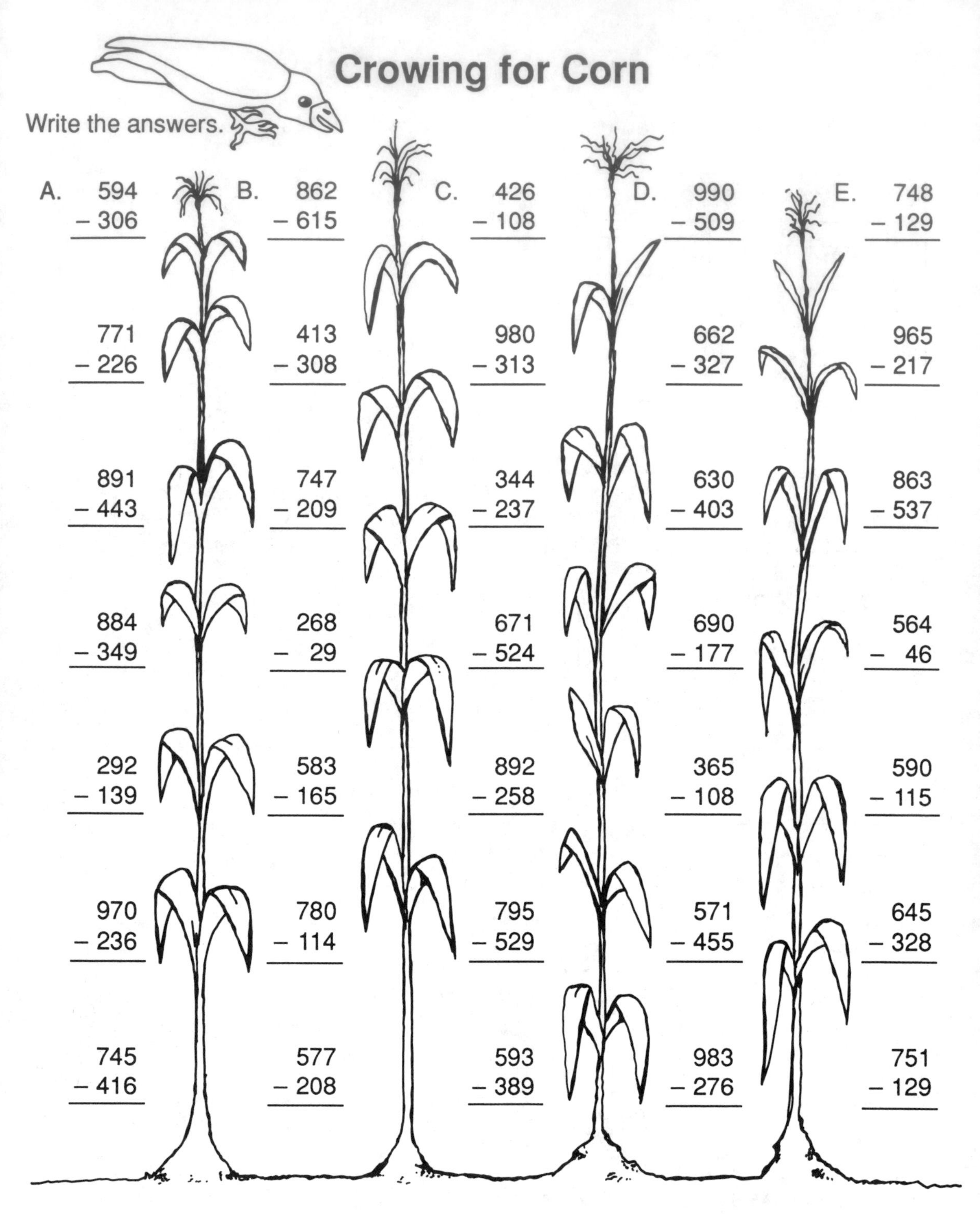

A.	B.	C.	D.	E.
594 − 306	862 − 615	426 − 108	990 − 509	748 − 129
771 − 226	413 − 308	980 − 313	662 − 327	965 − 217
891 − 443	747 − 209	344 − 237	630 − 403	863 − 537
884 − 349	268 − 29	671 − 524	690 − 177	564 − 46
292 − 139	583 − 165	892 − 258	365 − 108	590 − 115
970 − 236	780 − 114	795 − 529	571 − 455	645 − 328
745 − 416	577 − 208	593 − 389	983 − 276	751 − 129

Brainwork! Circle one of the problems above. Write a word problem to go with it.

Subtraction Balancing Act

Subtract and write the answers.

A.	737 − 450	956 − 895	638 − 147	968 − 486	928 − 345	526 − 336
B.	627 − 282	919 − 622	958 − 262	959 − 571	829 − 179	419 − 350
C.	814 − 362	628 − 390	375 − 193	837 − 762	637 − 481	847 − 464
D.	946 − 794	748 − 273	849 − 254	727 − 465	768 − 674	829 − 558
E.	768 − 398	989 − 197	478 − 181	619 − 546	437 − 296	847 − 683
F.	437 − 77	759 − 585	672 − 280	549 − 183	876 − 283	

Brainwork! Write a subtraction word problem about a circus seal. Solve the problem.

Funny Frog

First subtract and write the answers. Then use the key to write a letter below each problem. Read the riddle and its answer.

Key		
188	=	N
264	=	F
276	=	W
289	=	G
358	=	R
376	=	A
454	=	I
477	=	E
486	=	S
525	=	T
585	=	P
599	=	K
656	=	O
665	=	D
788	=	H

522 − 246	881 − 93	550 − 174	721 − 196
W			

943 − 278	842 − 186

851 − 587	617 − 259	955 − 299	488 − 199	782 − 296

933 − 268	610 − 252	630 − 176	666 − 478	797 − 198

741 − 365	820 − 295

754 − 169	762 − 386	627 − 269	912 − 387	651 − 197	610 − 133	683 − 197

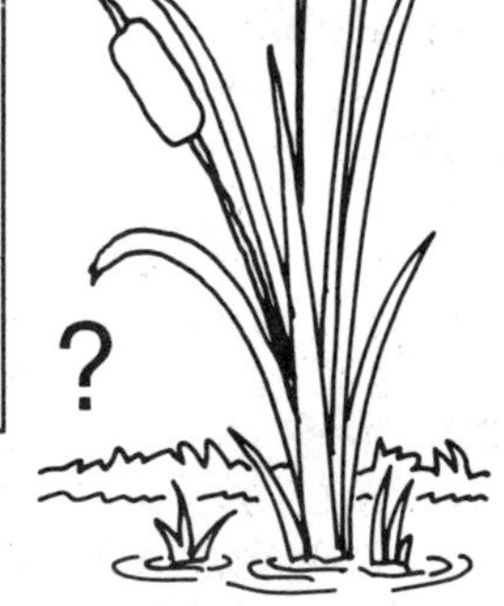

Answer: CROAKA-COLA!

Brainwork! Write subtraction problems and a matching key to spell out a name for the frog.

Pull-Out Answers

Page One

A. 12, 11, 10, 14, 9, 11
B. 12, 13, 10, 15, 8, 18
C. 14, 8, 10, 13, 12, 9
D. 9, 10, 14, 13, 13, 13
E. 14, 11, 15, 15, 11, 11
F. 10, 12, 12, 9, 12, 11
G. 16, 11, 10, 15, 11, 13
H. 10, 16, 12, 10
I. 17, 16, 14, 17

Page Two

		13 − 6 = 7	18 − 8 = 10	18 − 9 = 9		
	11 − 6 = 5	12 − 5 = 7	14 − 8 = 6	17 − 9 = 8	10 − 7 = 3	
13 − 4 = 9	12 − 9 = 3	15 − 8 = 7	7 − 4 = 3	12 − 4 = 8	16 − 7 = 9	13 − 8 = 5
11 − 8 = 3	16 − 9 = 7	10 − 4 = 6	13 − 7 = 6	11 − 3 = 8	14 − 9 = 5	8 − 3 = 5
14 − 7 = 7	12 − 3 = 9	12 − 8 = 4	15 − 6 = 9	10 − 9 = 1	9 − 2 = 7	10 − 3 = 7
17 − 8 = 9	15 − 7 = 8	13 − 5 = 8	11 − 4 = 7	9 − 4 = 5	12 − 7 = 5	8 − 5 = 3
START 14 − 6 = 8	11 − 7 = 4	14 − 5 = 9	9 − 6 = 3	10 − 5 = 5	7 − 2 = 5	11 − 2 = 9

Page Three

	hundreds	tens	ones	
A.	4	2	5	425
B.	6	3	9	639
C.	3	4	2	342
D.	1	7	8	178
E.	4	5	3	453

Try This! 639, 453, 425, 342, 178

Page Four

A. < E. < I. >
B. < F. > J. <
C. > G. > K. >
D. > H. < L. <

Try This! 130, 139, 150, 170, 200, 271, 289, 300, 394, 397, 450, 481, 483, 550, 640, 643, 650, 698, 700, 700, 800, 801, 810, 875

Page Five

A. b B. a
C. d D. c
E. 9,000 + 500 + 80 + 9
F. 5,000 + 200 + 60 + 6
G. 8,000 + 600 + 30 + 1
H. 3,000 + 0 + 70 + 5

Page Six

A. < I. >
B. < J. =
C. < K. <
D. > L. <
E. > M. =
F. > N. >
G. > O. =
H. =

Color the dinosaur **black.**

Page Seven

START →	41	85	71	65	49
78	98	88	60	64	91
76	62	56	89	81	73
57	61	73	49	92	80
59	70	76	66	60	74
79	61	75	69	77	82
80	48	54	46	95	

Page Eight

Picture should be colored according to the code.

Page Nine

A. 60, 58, 73, 79
B. 58, 76, 97, 91
C. 71, 59, 70, 64, 93
D. 55, 41, 67, 40
E. 73, 96, 61, 98, 37
F. 55, 77, 74, 38
G. 87, 57, 49, 62, 79
H. 77, 57, 43, 87
I. 48, 90, 95, 96, 84

Page Ten

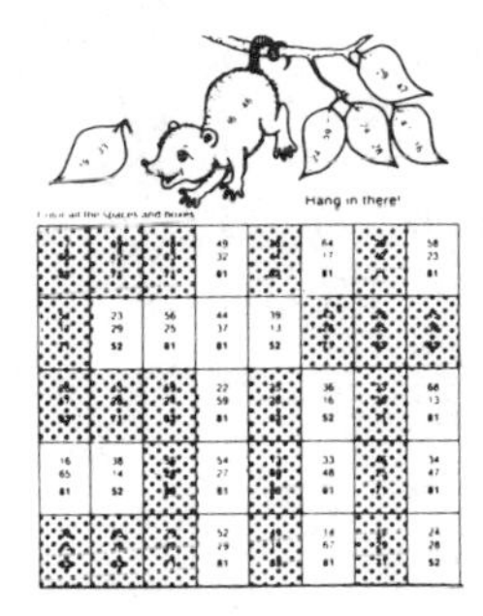

Page Eleven

A. 916, 501, 529, 819, 748, 947
B. 306, 807, 825, 828, 943, 649
C. 509, 714, 739, 739, 764, 609
D. 646, 918, 358, 902, 248, 818
E. 858, 968, 917, 506, 627, 334
F. 507, 729, 929, 826, 942, 927
G. 808, 819, 920, 734

Page Twelve

A. 762 F. 539 K. 812
B. 970 G. 969 L. 852
C. 842 H. 620 M. 585
D. 888 I. 922 N. 433
E. 960 J. 945 O. 888

Page Thirteen

	START →	974	791	773	669
491	1,229	1,154	848	526	748
995	589	1,257	737	1,049	518
671	454	1,396	1,622	530	1,056
1,815	818	985	633	703	305
895	1,214	662	482	866	1,125
1,481	777	1,027	1,543	1,270	365

Spot will use doghouse E.

Pull-Out Answers

Page Fourteen

A. 660; 940; 922; 1,013; 1,214; 722; 512
B. 931; 810; 1,280; 1,363; 521; 842; 750
C. 907; 1,058; 551; 1,532; 342; 517; 733
D. 600; 1,344; 498; 300; 1,000; 950; 695
E. 702; 221; 1,504; 742; 1,172; 1,201; 800

Page Fifteen

A. 399; 1,241; 888; 1,216; 237; 777
B. 495; 736; 736; 586; 944; 1,657
C. 876; 595; 249; 1,188; 461; 524
D. 714; 476; 1,037; 1,097; 590; 698
E. 776; 1,330; 578; 872; 345; 849

Page Sixteen

A. 5,995; 9,400; 6,513; 3,689; 7,168
B. 4,552; 3,080; 9,497; 9,148; 1,806
C. 8,436; 5,226; 4,311; 9,233; 3,897
D. 2,221; 9,829; 6,490; 1,793; 5,576
E. 4,896; 2,873; 7,690; 2,958; 8,000
F. 5,541; 3,529; 9,472; 6,668; 7,879
G. 2,849; 4,225; 5,456; 9,906; 7,185

Page Seventeen

A. 38, 25, 69, 27, 46, 45
B. 43, 72, 55, 17, 9, 32
C. 59, 6, 61, 34, 47, 26
D. 44, 19, 36, 9, 39, 84
E. 47, 64, 75, 12, 67, 78

Page Eighteen

A. 15	J. 53
B. 12	K. 19
C. 16	L. 7
D. 48	M. 26
E. 9	N. 46
F. 23	O. 65
G. 28	P. 38
H. 47	Q. 15
I. 57	R. 11

Page Nineteen

		21	85	38	48
54	28	50	25	78	56
52	36	76	65	44	18
25	17	3	51	58	62
45	76	48	64	38	54
61	7	15	8	48	52
58	16	33	72	26	54

Petunia won 3rd prize.

Page Twenty

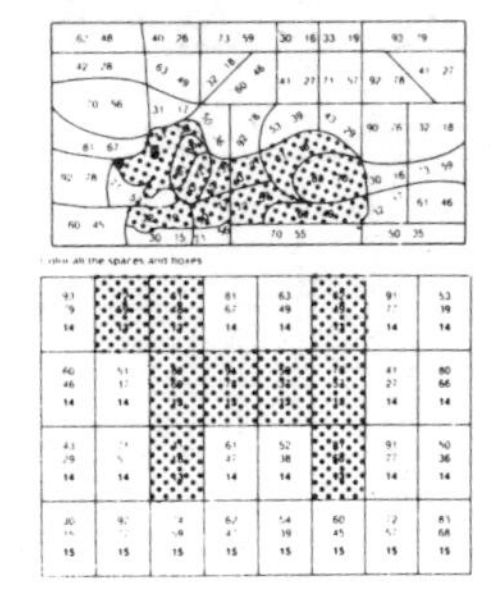

Page Twenty-one

214 623 268 268 239 619
R A B B I T
239 277
I S
595 214 122 384 446
P R O U D
122 388 253 122 384
O F Y O U

Try This! 119

Page Twenty-two

A.	B.	C.
288	247	318
545	105	667
448	538	107
535	239	147
153	418	634
734	666	266
329	369	204

D.	E.
481	619
335	748
227	326
513	518
257	475
116	317
707	622

Page Twenty-three

A. 287, 61, 491, 482, 583, 190
B. 345, 297, 696, 388, 650, 69
C. 452, 238, 182, 75, 156, 383
D. 152, 475, 595, 262, 94, 271
E. 370, 792, 297, 73, 141, 164
F. 360, 174, 392, 366, 593

Page Twenty-four

276 788 376 525 665 656
W H A T D O
264 358 656 289 486
F R O G S
665 358 454 188 599 376 525
D R I N K A T
585 376 358 525 454 477 486
P A R T I E S ?

Pull-Out Answers

Page Twenty-five

A. 4,138; 2,470; 2,203;
5,534; 6,304; 3,050

B. 589; 3,939; 1,308; 1,599;
1,224; 7,813

C. 3,417; 5,365; 667; 5,850;
2,087; 6,729

D. 6,154; 3,006; 2,752;
6,664; 5,566; 2,345

E. 2,419; 3,234; 6,347; 478;
7,613; 3,080

F. 5,697; 1,788; 4,438;
7,277; 5,182

Page Twenty-six

1. quarter past five, 5:15
2. half past two, 2:30
3. quarter to four, 3:45
4. quarter past eleven, 11:15
5. half past ten, 10:30
6. quarter past six, 6:15
7. quarter to nine, 8:45
8. half past seven, 7:30
9. quarter past four, 4:15
10. quarter to two, 1:45
11. half past twelve, 12:30
12. quarter to eleven, 10:45

Page Twenty-seven

A. 3:00; 6:15
3 hours and **15** minutes

B. 7:00; 9:20
2 hours and **20** minutes

C. 10:00; 12:10
2 hours and **10** minutes

D. 11:00; 5:30
6 hours and **30** minutes

E. 4:00; 8:15
4 hours and **15** minutes

F. 1:00; 9:05
8 hours and **5** minutes

G. 2:00; 7:45
5 hours and **45** minutes

Try This! 75 minutes

Page Twenty-eight

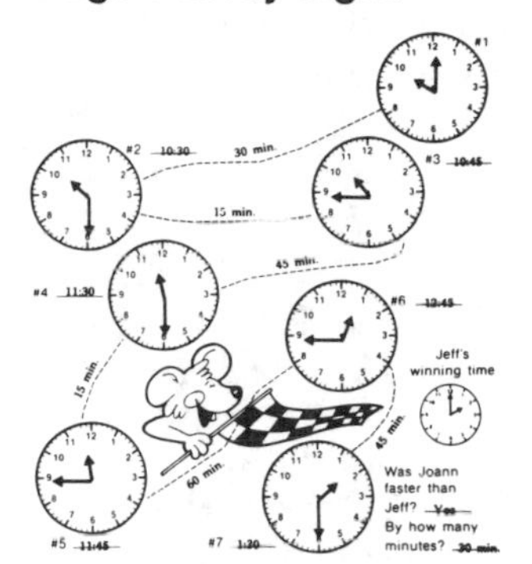

Page Twenty-nine

1. $1.63
2. $1.15
3. $1.68
4. $1.44
5. $1.37
6. $1.58
7. $1.65
8. $1.84

Brainwork! $1.15, $1.37,
$1.44, $1.58, $1.63,
$1.65, $1.68, $1.84

Page Thirty

B. Total = $1.05
Change = $.20

C. Total = $3.55
Change = $.20

D. Total = $1.42
Change = $.08

E. Total = $1.02
Change = $.98

Page Thirty-one

A. $.31
☐ yes

B. $.27
☐ yes

C. $.35
☐ no

D. $.70
☐ no

E. $.85
☐ yes

F. $.38
☐ no

G. $.13
☐ yes

E. $.60
☐ yes

Page Thirty-two

1. apples
2. 2
3. 60¢
4. 8
5. 40¢
6. grapefruit

Page Thirty-three

1. cheetah
2. 50 m.p.h.
3. grizzly bear, house cat
4. zebra
5. rabbit
6. 15 m.p.h.
7. rabbit
8. 20 m.p.h.

Page Thirty-four

1. 40 minutes
2. Monday, Tuesday,
and Thursday
3. 20 minutes
4. Saturday
5. Sunday
6. 160 minutes

Page Thirty-five

Picture should be colored according to the code.

Page Thirty-six

1. 36, 35, 4, 0, 32
2. 30, 16, 15, 8, 25
3. 12, 40, 24, 0, 45
4. 10, 20, 5, 28, 20

Page Thirty-seven

36	42
0	56
63	30
48	28
35	18
6	42
49	0
24	14
21	54
12	7

Pull-Out Answers

Page Thirty-eight

A = 0 I = 24 U = 63 P = 64
O = 8 C = 48 M = 45 L = 40
T = 18 N = 72

MULTIPLICATION TABLE

H = 0 I = 72 D = 16 Y = 36
R = 56 N = 9 T = 32 C = 27
I = 72 E = 81

IN THE DICTIONARY

Page Thirty-nine

Picture should be colored according to the code.

Page Forty

AN ELEPHANT'S SHADOW

Page Forty-one

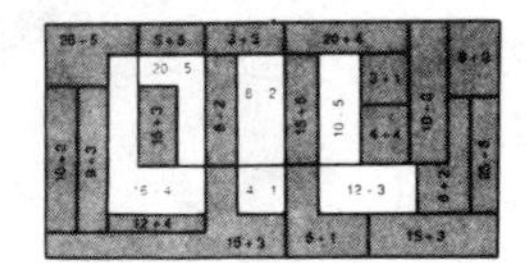

Page Forty-two

1. 18 ÷ 3 = 6, 6
2. 36 ÷ 4 = 9, 9
3. 24 ÷ 6 = 4, 4
4. 16 ÷ 8 = 2, 2
5. 21 ÷ 3 = 7, 7
6. 32 ÷ 4 = 8, 8
7. 27 ÷ 3 = 9, 9

Brainwork! 18 ÷ 3 = 6, 6 days

Page Forty-three

A NIGHT WATCHMAN

Row 1: T)9 A)2 B)1 M)15
Row 2: R)6 E)5 C)16 S)8
Row 3: Z)10 G)12 L)4 H)20
Row 4: I)14 W)3 N)18 L)0

Page Forty-four

1–5. 4, 6, 3, 2, 0
6–10. 3, 3, 16, 1, 12
11–15. 5, 3, 3, 12, 8
16–20. 4, 25, 8, 5, 2
21–25. 15, 5, 4, 6, 2
26–30. 3, 4, 5, 2, 10
31–33. 3, 4, 4
34–36. 3, 5, 20

Page Forty-five

A. 4/5, 1/5
B. 2/3, 1/3
C. 5/8, 3/8
D. 1/4, 3/4
E. 9/10, 1/10
F. 1/2, 1/2

Try This! The same amount of pie has been eaten.

Page Forty-six

A. 1/3 B. 4/6
C. 5/8 D. 2/4
E. 2/5 F. 1/2

Try This!

A. $1/3 < 2/3$ B. $4/6 < 5/6$
C. $7/8 > 5/8$ D. $2/4 < 3/4$
E. $3/5 > 2/5$ F. $4/6 > 1/2$

Page Forty-seven

1. D A. square
2. A B. hexagon
3. C C. rectangle
4. F D. triangle
5. B E. circle
6. E F. pentagon

Page Forty-eight

1. 2 spheres
2. 4 cones
3. 6 cubes
4. 4 cylinders
5. 4 rectangular prisms

Coiled for Math

Subtract and write the answers.

A.	8,283 − 4,145	6,660 − 4,190	3,382 − 1,179	7,252 − 1,718	8,619 2,315	4,782 − 1,732
B.	3,204 − 2,695	6,986 − 3,047	1,434 − 126	8,188 − 6,589	2,318 − 1,094	8,964 − 1,151
C.	6,494 − 3,077	8,545 − 3,180	8,701 − 8,034	7,040 − 1,190	7,365 − 5,278	7,782 − 1,053
D.	8,643 − 2,489	5,913 − 2,907	4,915 − 2,163	8,987 − 2,323	8,162 − 2,596	3,456 − 1,111
E.	4,303 − 1,884	5,421 − 2,187	9,682 − 3,335	9,481 − 9,003	9,949 − 2,336	4,652 − 1,572
F.	8,919 − 3,222	9,413 − 7,625	5,556 − 1,118	8,403 − 1,126	9,829 − 4,647	

Write the time shown on the clock: (1) quarter to; quarter past; half past.
(2) Write the time using numbers. For example: quarter to ten - 9:45

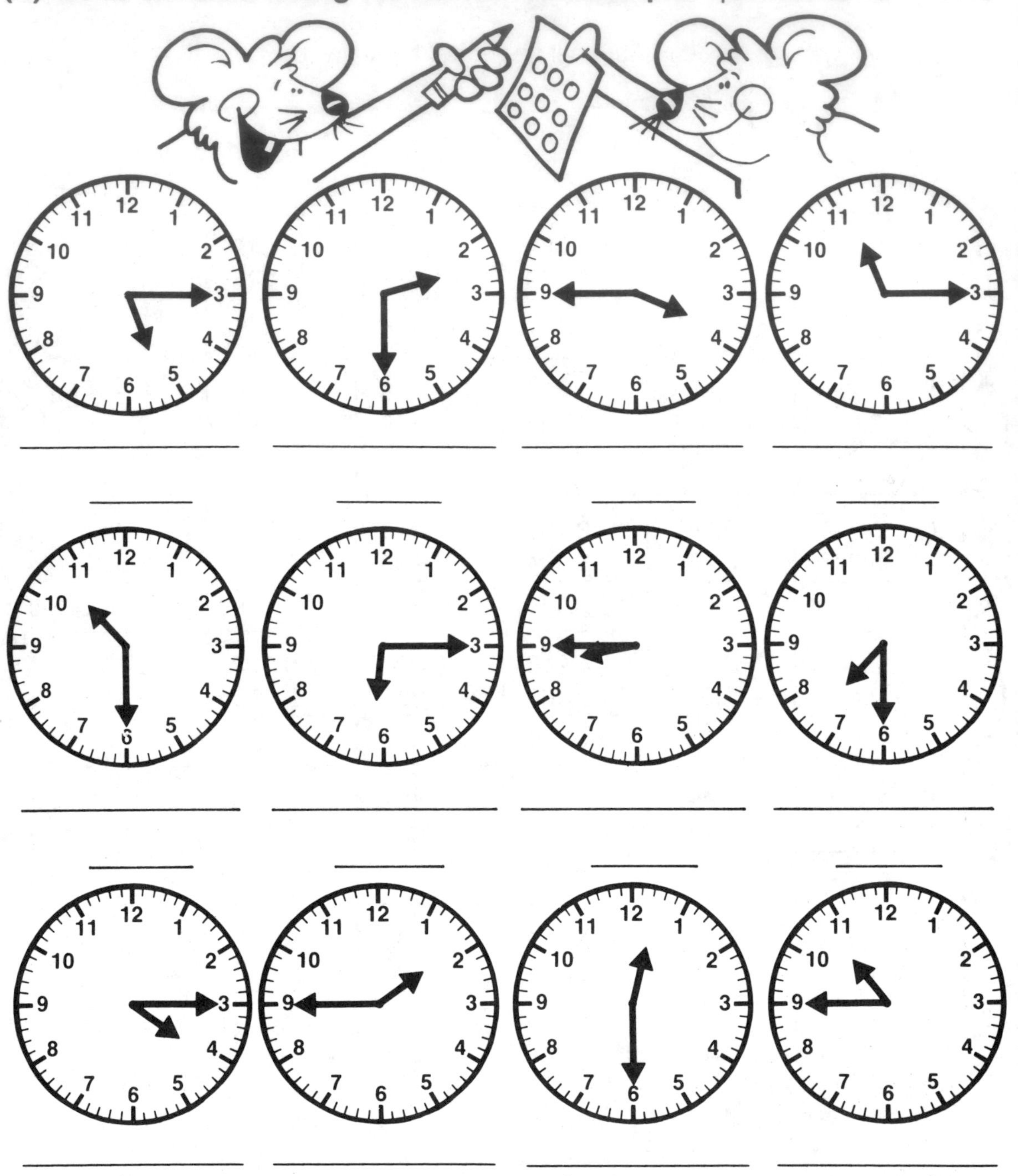

Hours and Minutes

Write the time shown on each clock. For each pair decide how much time has passed. Write the number of hours and minutes.

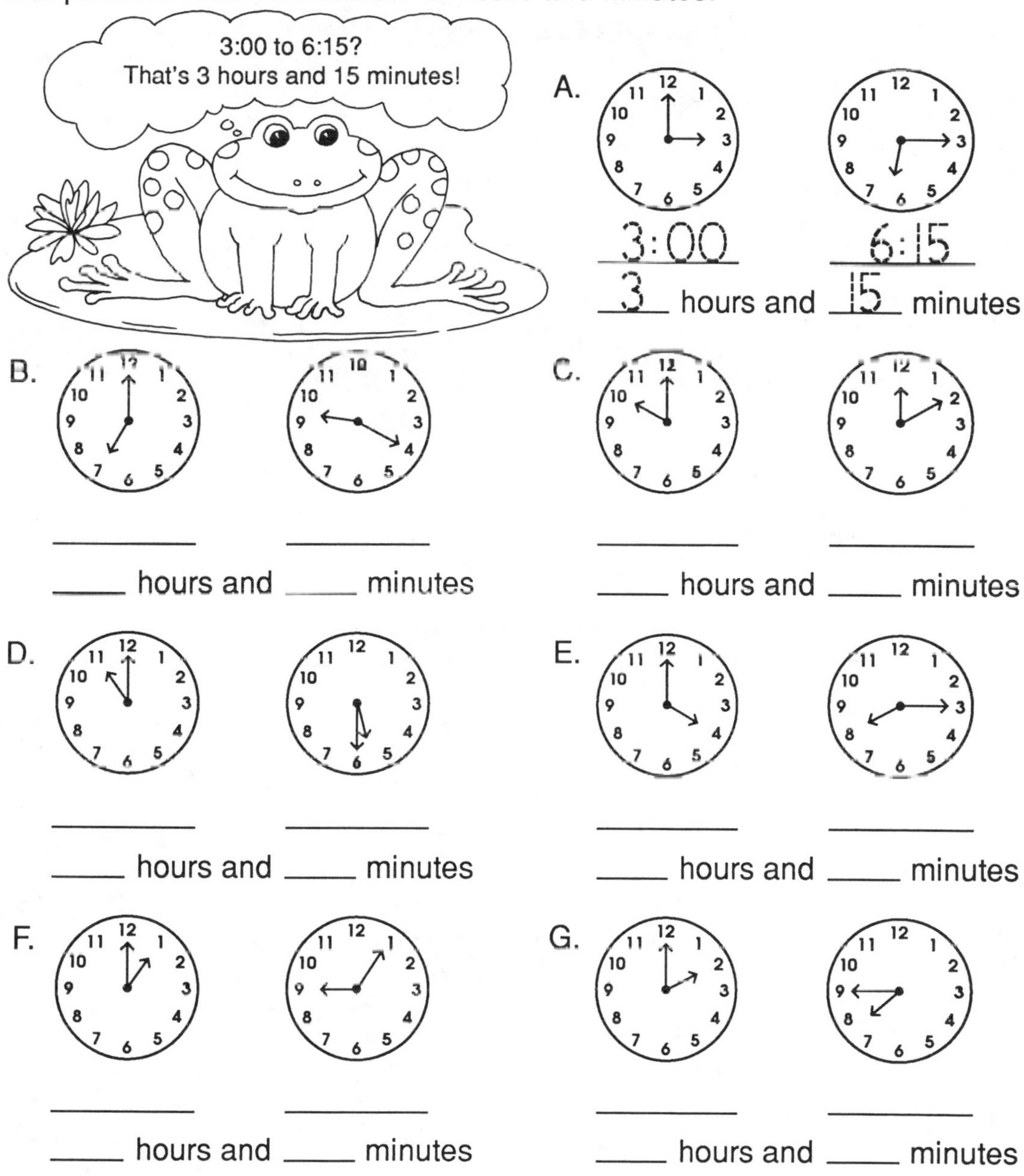

Try This! One hour is 60 minutes, so one hour and 30 minutes is 90 minutes. Write how much time has passed in minutes from 3:00 to 4:15.

"The big race is tomorrow!" exclaimed Joann. "I'm going to win this year. My bike is the fastest!"

At the starting line, a man said, "Last year, Jeff Nelsen won this race. It took him three hours and forty-five minutes to finish. Will someone be able to beat that time?" Joann got ready. The man waved the flag and she was off!

Put your pencil on clock #1. Draw the hands to 10 a.m. Follow the road with Joann to the next clock. It took thirty minutes to get there. What time is it now? Write the time on the line. Draw the hands on the clock. Follow the road to the end.

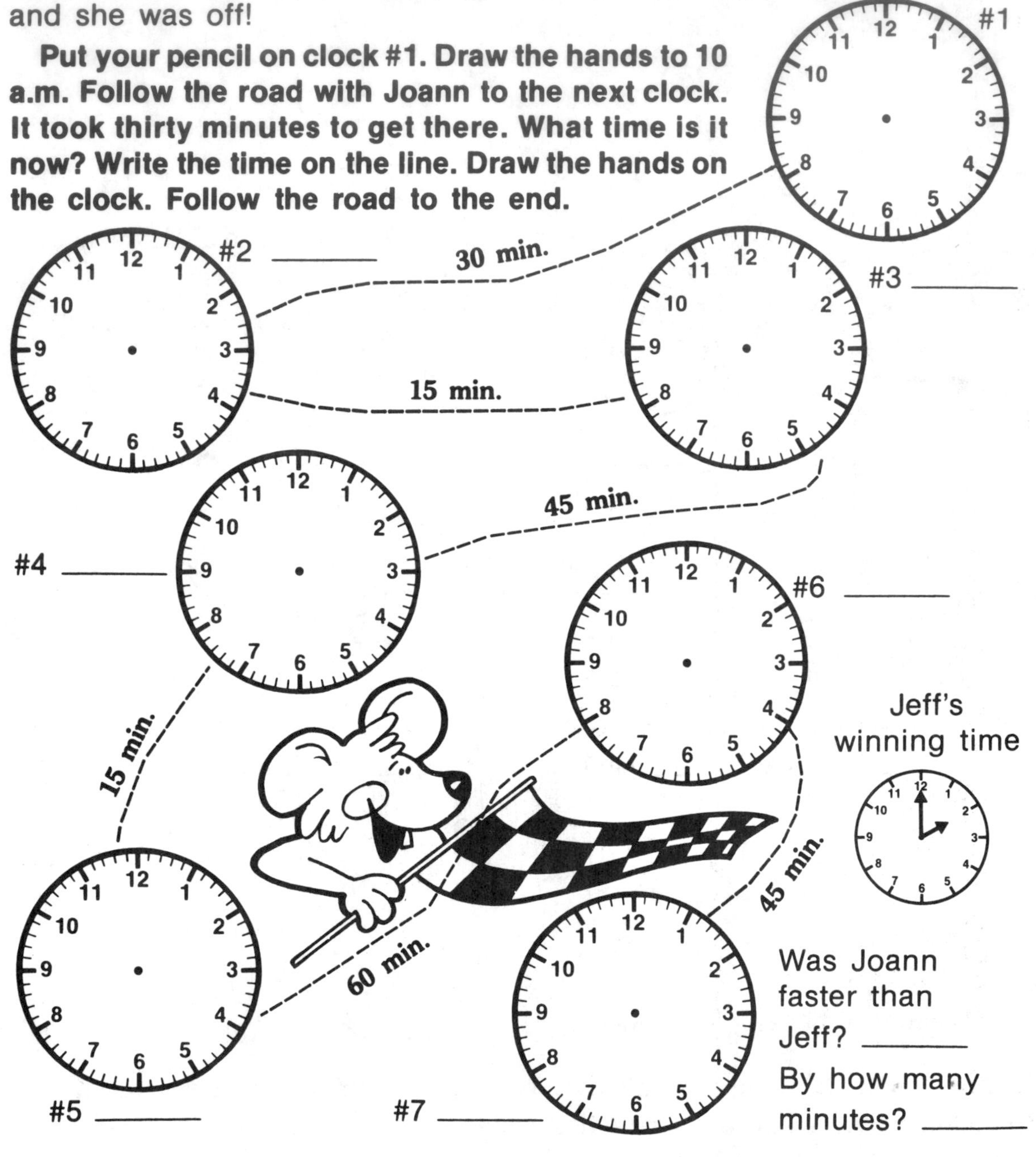

Dollars and Cents

How much money is in each set?
Use a dollar sign and a cents point as you write the amounts.

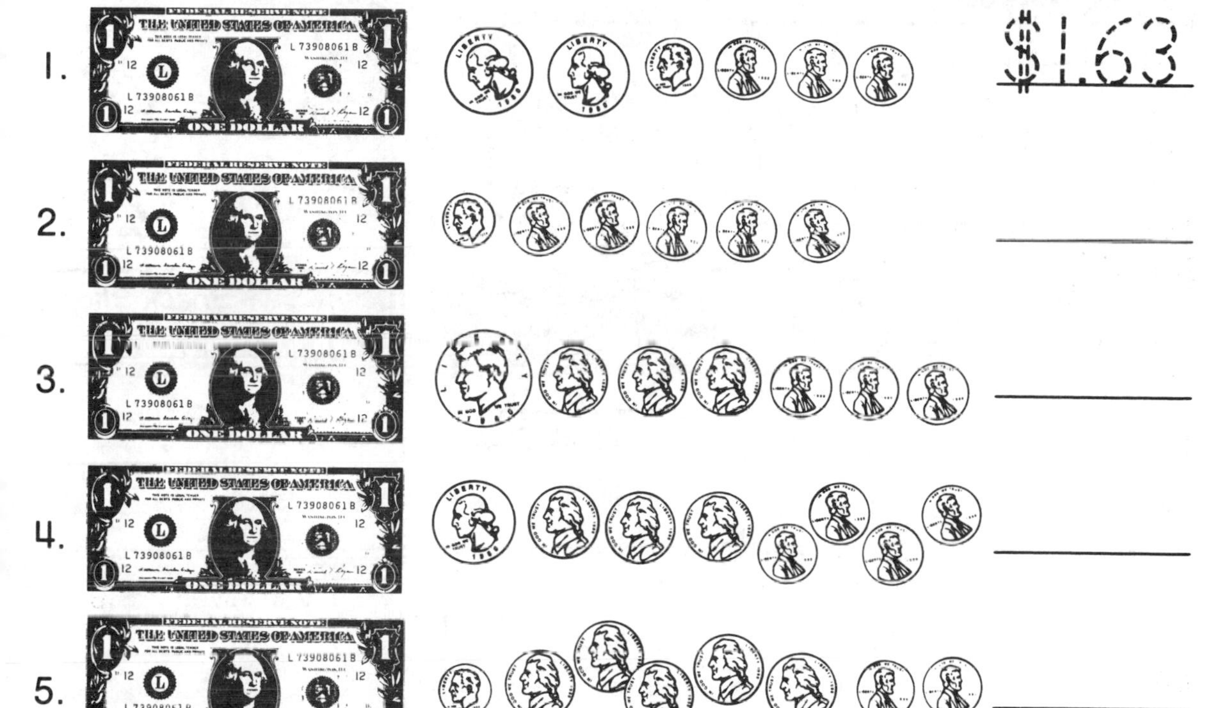

Brainwork! Write the amounts in order from the lowest to the highest in value.

The Snack Shop

The Snack Shop needs your help. The cash registers are not working. Fill in the prices and total up the orders. Be sure to give back the correct change. Order **A** has been done for you.

Menu

Superburger	$.89	Green Salad	$.39
Cheeseburger	$.58	Milk	$.25
Pizza Slice	$.49	Apple Juice	$.55
Fruit Cup	$.28	Yogurt	$.35

A. ***Snack Shop***

1 Yogurt	$.35
1 Milk	$.25
1 Fruit Cup	$.28
Total Cost	$.88
Money received	$.95
Total Cost	–$.88
Change	$.07

B. ***Snack Shop***

3 Yogurts	______
	______
	______
Total Cost	______
Money received	$1.25
Total Cost	______
Change	______

C. ***Snack Shop***

2 Superburgers	______
	______
1 Fruit Cup	______
1 Green Salad	______
2 Apple Juice	______
	______
Total Cost	______
Money received	$3.75
Total Cost	______
Change	______

D. ***Snack Shop***

1 Superburger	______
1 Milk	______
1 Fruit Cup	______
Total Cost	______
Money received	$1.50
Total Cost	______
Change	______

E. ***Snack Shop***

1 Pizza Slice	______
1 Milk	______
1 Fruit Cup	______
Total Cost	______
Money received	$2.00
Total Cost	______
Change	______

Brainwork! Order lunch for yourself. Total the amount.

Making Change for a Dollar

Is the change from a dollar correct?
Read the cost of each item. Write the amount of change shown in each coin purse.
Mark the box ☒ yes if the amount of change is correct, ☒ no if it is not correct.

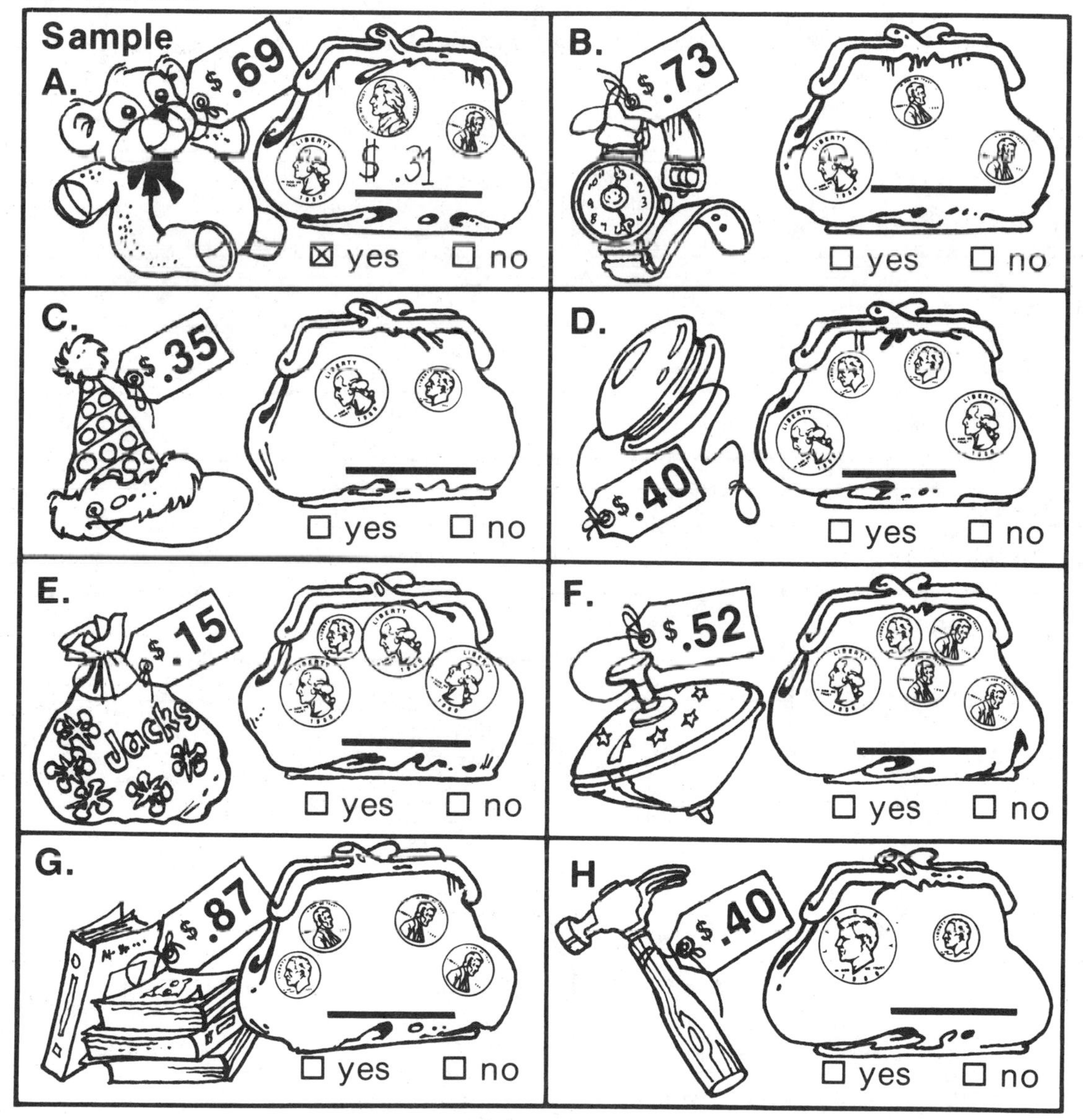

Try This! Write three interesting problems about making change for $5. Use a robot buying replacement parts.

Snack Sale

Mr. Jones's third grade class sold snacks at the fair to raise money for a class party.

The picture graph shows the number of snacks that were sold.

Use it to answer the questions. ☆ = 2 snacks sold

apples	☆ ☆ ☆ ☆ ☆ ☆
grapefruit	☆
bananas	☆ ☆ ☆ ☆
peaches	☆ ☆ ☆
pears	☆ ☆
oranges	☆ ☆ ☆

1. Which snack sold best? ____________________
2. How many more oranges were sold than pears? ____________
3. Each peach sold for 10¢. How much money did the class make selling peaches? __________________
4. How many bananas were sold? ________________
5. Each banana sold for 5¢. How much money did the class make selling bananas? ______________
6. Which snack did not sell well? __________________

Brainwork! What is your favorite snack? Write a sentence telling why.

Speeds of Animals

Here is a bar graph that shows the speeds of some animals. Use the graph to answer the questions below.

Miles Per Hour (m.p.h.)

75
70
65
60
55
50
45
40
35
30
25
20
15
10
5
0

Cheetah
Lion
Zebra
Rabbit
Grizzly Bear
House Cat
Elephant
Wild Turkey

1. Which animal can run 70 m.p.h.? __________
2. How fast can a lion run? __________
3. Which two animals can run the same speed? __________ __________
4. Which is faster, a grizzly bear or a zebra? __________
5. Which of these animals is the slowest— a lion, zebra, cheetah, or rabbit? __________
6. How much faster is a zebra than an elephant? __________
7. A cheetah can run twice as fast as which animal? __________
8. How much slower is a house cat than a lion? __________

Ashley's Homework Time

A line graph can show changes over time. Lines on the graph can move up, down, or stay the same.

Mr. Black asked students in his class to record the amount of time they spent doing homework in one week.

Study Ashley's line graph and answer the questions below.

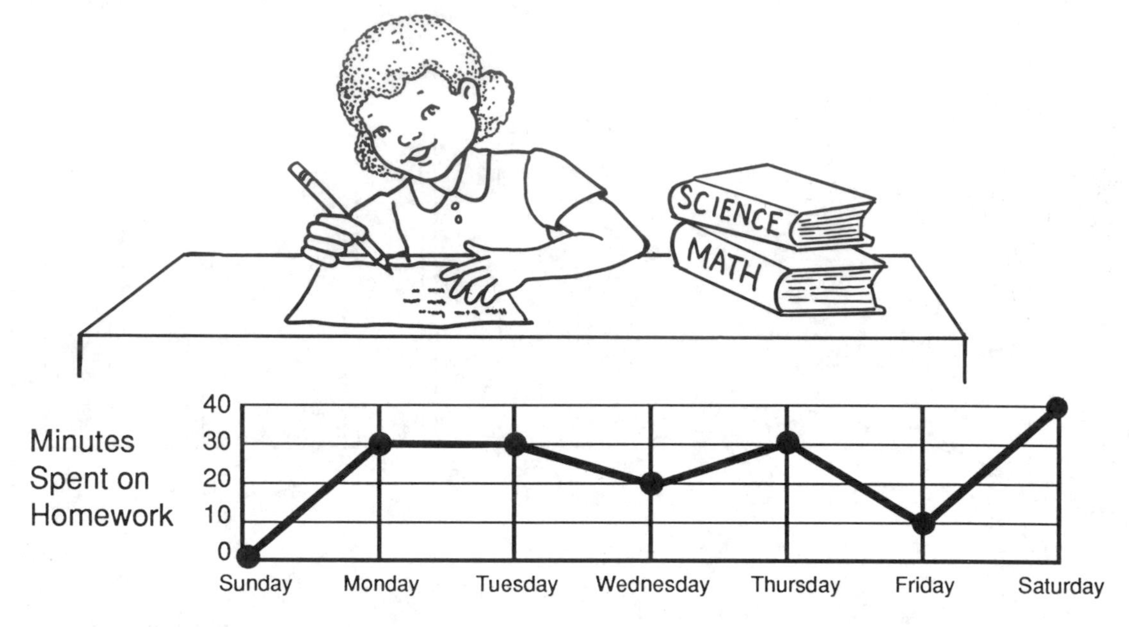

1. How many minutes did Ashley spend doing homework on Saturday?

   ____

2. On ____________, ____________,

   and ____________________,
   Ashley spent the same amount of time doing homework.

3. Ashley spent ______ minutes working on her homework on Wednesday.

4. On which day did she spend the most time doing homework?

   ____

5. On which day did Ashley spend the least time doing homework?

   ____

6. What was the total amount of minutes she spent doing her homework in one week?

   ____

Brainwork! Make your own line graph like the one above. Record the amount of time that you spend doing homework.

12	18	20	24	30
black	blue	brown	yellow	orange

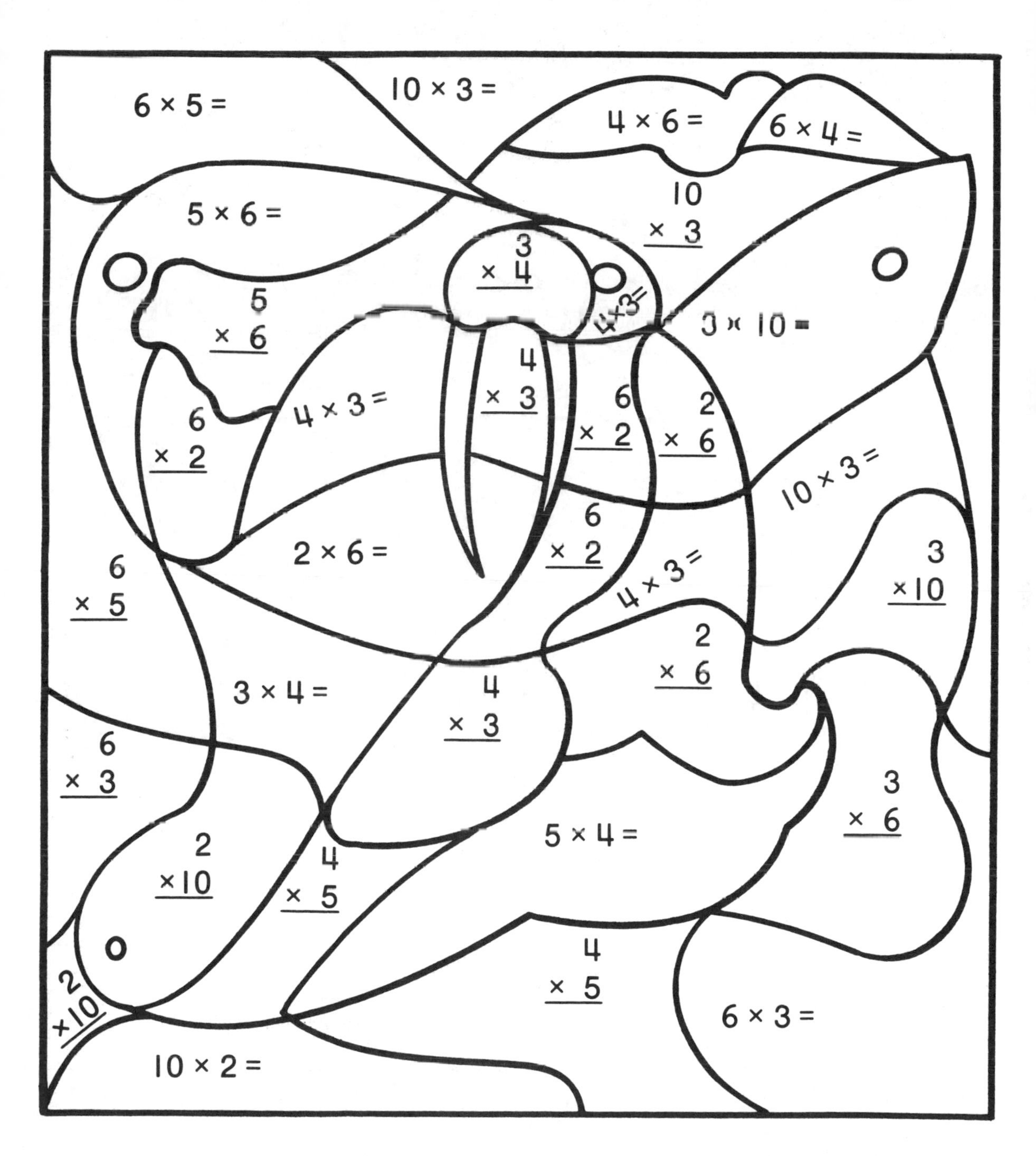

Don't Clown Around!

Color Code for Hats

Answer	8							4
	30	28	35	12	25	40	0	32
	36	24	10	45	20	15	5	16
Color	red	blue	green	orange	purple	black	brown	yellow

Circle the Correct Answer

Knock, Knock
Who's there?
Canoe
Canoe who?

6 × 6 =	33
	36
7 × 0 =	7
	0
7 × 9 =	61
	63
6 × 8 =	48
	40
7 × 5 =	30
	35
6 × 1 =	7
	6
7 × 7 =	48
	49
6 × 4 =	23
	24
7 × 3 =	21
	31
6 × 2 =	12
	8

6 × 7 =	42
	45
7 × 8 =	63
	56
6 × 5 =	30
	20
7 × 4 =	28
	21
6 × 3 =	24
	18
7 × 6 =	41
	42
6 × 0 =	0
	6
7 × 2 =	14
	18
6 × 9 =	54
	42
7 × 1 =	0
	7

Answer: Canoe come out and play?

Riddle: What table has no legs?

A 9×0	I 8×3	U 9×7	P 8×8	O 8×1
C 8×6	M 9×5	L 8×5	T 9×2	N 9×8

Fill in the correct letter over each answer.

Answer: ___ ___ ___ ___ ___ ___ **L I** ___ ___ **T I** ___ ___

45 63 40 18 24 64 48 0 8 72

TABLE

Riddle: Where can you always find money when you look for it?

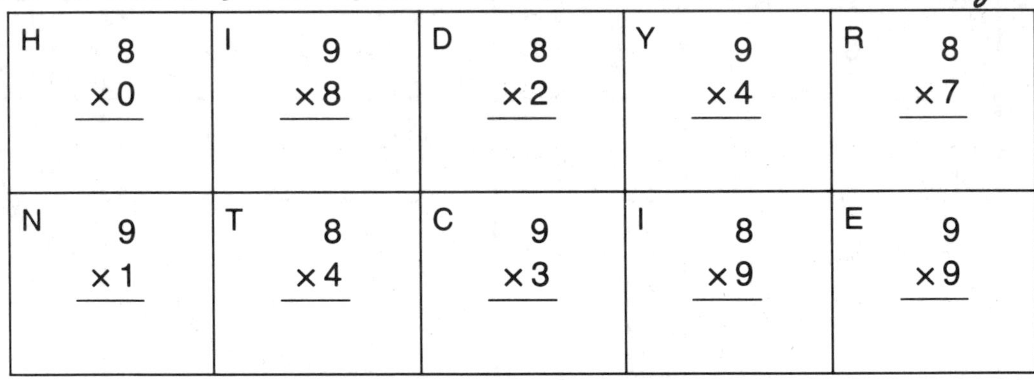

H 8×0	I 9×8	D 8×2	Y 9×4	R 8×7
N 9×1	T 8×4	C 9×3	I 8×9	E 9×9

Fill in the correct letter over each answer.

Answer: I N T ___ ___

0 81

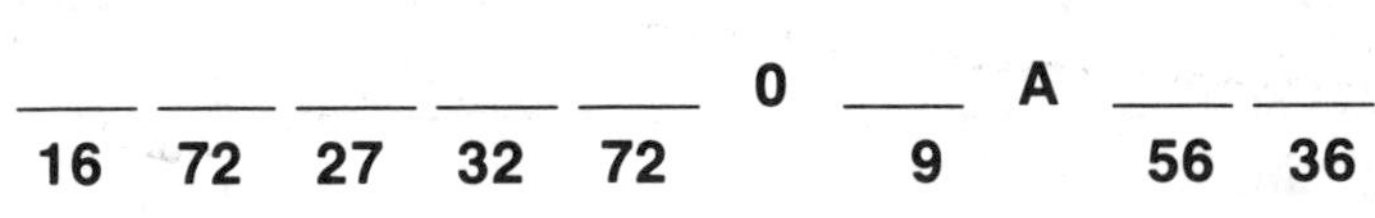

___ ___ ___ ___ ___ **O** ___ **A** ___ ___

16 72 27 32 72 9 56 36

40 green

42 red

54 brown

56 blue

63 orange

$9 \times 7 =$

$6 \times 7 =$

$7 \times 6 =$

$\begin{array}{r} 9 \\ \times\ 7 \\ \hline \end{array}$

$\begin{array}{r} 7 \\ \times\ 9 \\ \hline \end{array}$

$\begin{array}{r} 8 \\ \times\ 5 \\ \hline \end{array}$

$\begin{array}{r} 5 \\ \times 8 \\ \hline \end{array}$

$7 \times 6 =$

$\begin{array}{r} 7 \\ \times\ 6 \\ \hline \end{array}$

$\begin{array}{r} 6 \\ \times 7 \\ \hline \end{array}$

$7 \times 9 =$

$7{\times}9 =$

$7{\times}6 =$

$\begin{array}{r} 9 \\ \times \\ \hline \end{array}$

$8 \times 5 =$

$\begin{array}{r} 7 \\ \times\ 8 \\ \hline \end{array}$

$7 \times 8 =$

$\begin{array}{r} 8 \\ \times 7 \\ \hline \end{array}$

$7{\times}8 =$

$\begin{array}{r} 8 \\ \times\ 7 \\ \hline \end{array}$

$8 \times 7 =$

$\begin{array}{r} 6 \\ \times 9 \\ \hline \end{array}$

$\begin{array}{r} 10 \\ \times\ 4 \\ \hline \end{array}$

$4 \times 10 =$

$\begin{array}{r} 9 \\ \times 6 \\ \hline \end{array}$

$\begin{array}{r} 9 \\ \times\ 6 \\ \hline \end{array}$

$\begin{array}{r} 7 \\ \times\ 8 \\ \hline \end{array}$

$6 \times 9 =$

$8 \times 7 =$

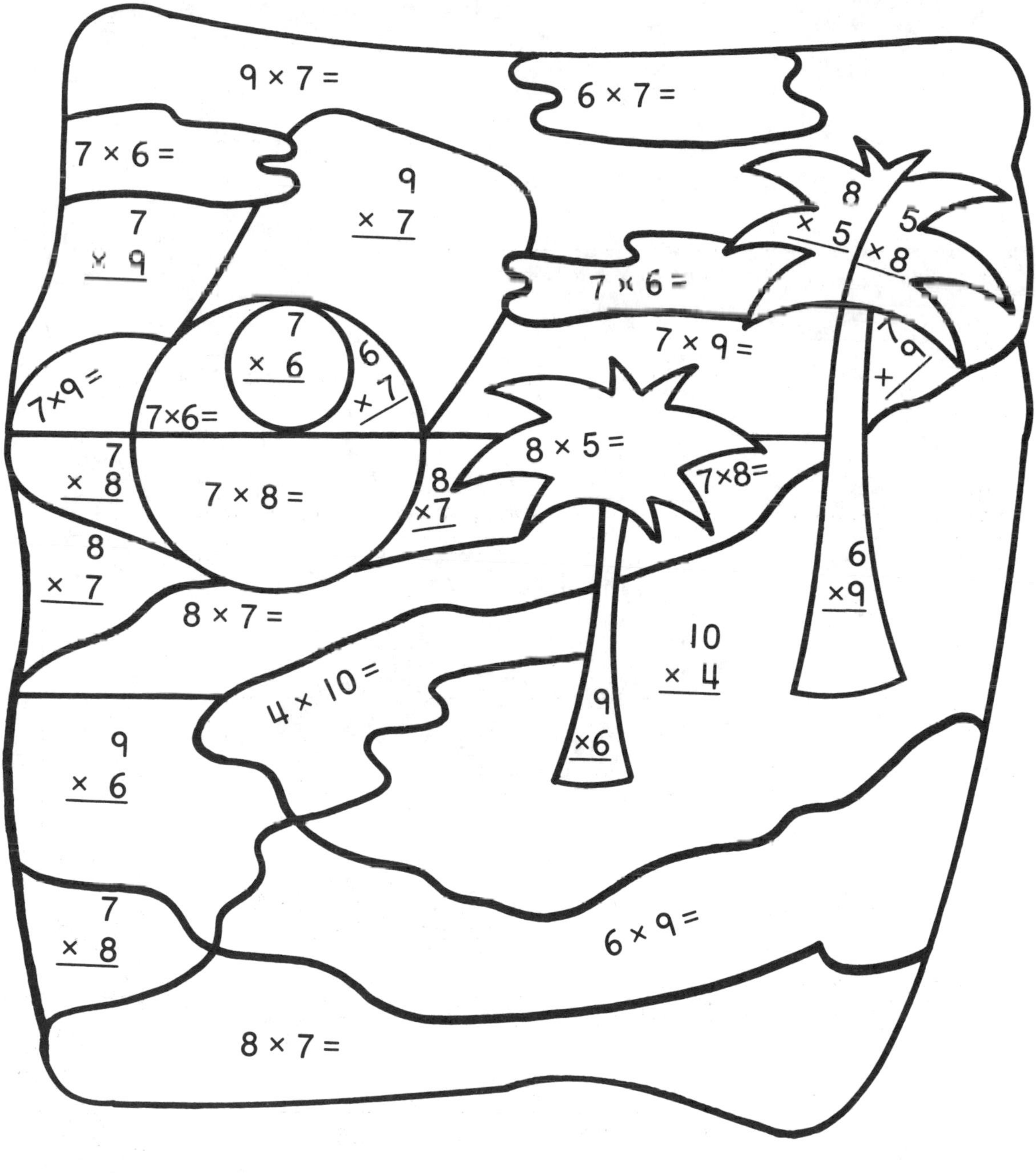

Riddle Fun

What is as big as an elephant but doesn't weigh anything?

Find the answers to the problems below.

$16 \div 2 =$ ________ E

$12 \div 3 =$ ________ N

$9 \div 3 =$ ________ T

$14 \div 2 =$ ________ D

$18 \div 9 =$ ________ P

$18 \div 3 =$ ________ S

$27 \div 3 =$ ________ O

$2 \div 2 =$ ________ H

$10 \div 2 =$ ________ A

$33 \div 3 =$ ________ L

$10 \div 1 =$ ________ W

Write the letters following each answer on the lines to solve the riddle.

___ ___   ___ ___ ___ ___ ___ ___ ___ ___ ' ___

5 4   8 11 8 2 1 5 4 3   6

___ ___ ___ ___ ___ ___

6 1 5 7 9 10

Thirsty?

What's a robot's favorite drink? To find its picture, solve each division problem. Color spaces with answers 1, 3, or 5 **green** and spaces with answers 2 or 4 **yellow**.

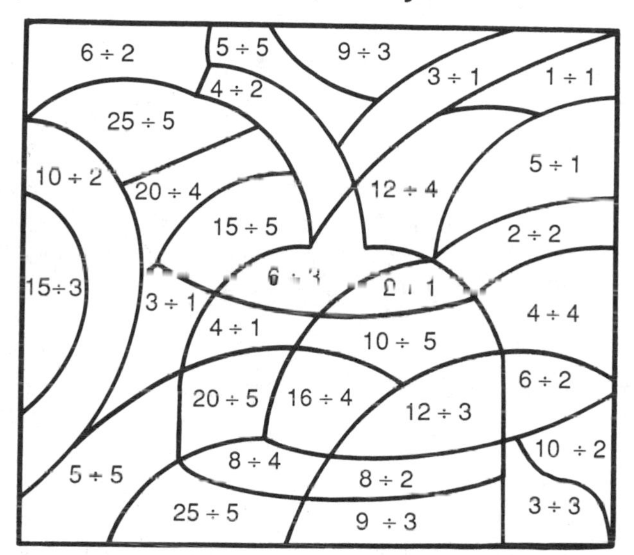

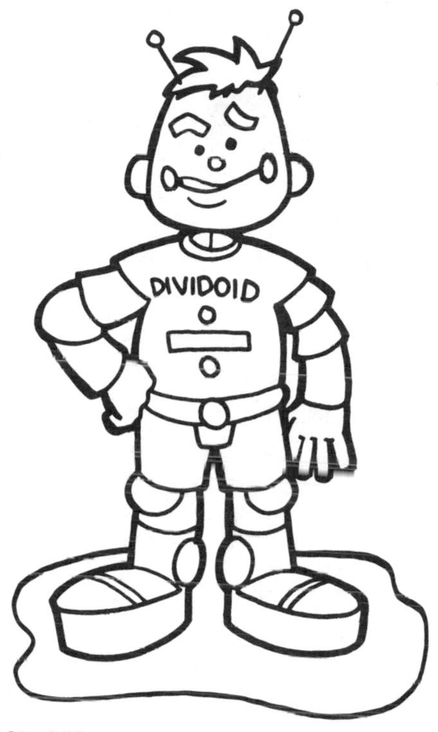

To find the drink's name, color spaces with answers 1, 3, or 5 **blue** and spaces with answers 2 or 4 **red**.

25 ÷ 5
5 ÷ 5
3 ÷ 3
20 ÷ 4
9 ÷ 3
20 ÷ 5
3 ÷ 1
8 ÷ 2
15 ÷ 3
6 ÷ 2
15 ÷ 5
10 ÷ 5
4 ÷ 4
10 ÷ 2
10 ÷ 2
9 ÷ 3
25 ÷ 5
16 ÷ 4
4 ÷ 1
12 ÷ 3
6 ÷ 2
12 ÷ 4
15 ÷ 3
5 ÷ 1
15 ÷ 3

Try This! You can write a division problem two different ways. For example, 25 ÷ 5 or $5\overline{)25}$. Write three other division facts two different ways.

In the Gym

Divide to solve these problems. Write the number sentence and the answer.

1. 18 children lined up in 3 rows to do situps. How many children were in each row?

   ____

   _______ children were in each row.

2. Ms. Simms divided 36 children into 4 teams. How many children were on each team?

   ____

   _______ children were on each team.

3. There are 6 tumbling mats. Ms. Simms divided 24 children into equal groups for each mat. How many children were in each group?

   ____

   _______ children were in each group.

4. 16 girls want to shoot baskets. Mr. Young will have the same number of girls at all 8 baskets. How many girls will be at each basket?

   ____

   _______ girls will be at each basket.

5. 3 equal groups of children want to run relay races. There are 21 children. How many children are in each group?

   ____

   _______ children are in each group.

6. 32 boys lined up in 4 equal rows to do jumping jacks. How many boys were in each row?

   ____

   _______ boys were in each row.

7. At the end of class, 27 kickballs must be put equally into 3 bags. How many kickballs fit in each bag?

   ____

   _______ kickballs fit in each bag.

Brainwork! Tom's dog eats 3 bones each day. How long will a box of 18 bones last? Draw pictures to prove your answer.

Stargazer

What's another name for an astronomer?

To find out, first solve each problem. Then write the letter of the problem on the line above its matching answer.

__   __ __ __ __ __   __ __ __ __ __ __ __ __

2   18 14 12 20 9   3 2 9 16 20 15 2 18

(T) 3×3

(A) $5\overline{)10}$

(B) $2\overline{)2}$

(M) 5×3

(R) $2 \times 3 =$

(E) $25 \div 5 =$

(C) $4 \times 4 =$

(S) $4 \times 2 =$

(Z) 5×2

(G) 3×4

(L) $4\overline{)16}$

(H) 4×5

(I) $2 \times 7 =$

(W) $12 \div 4 =$

(N) 3×6

(L) $0 \times 5 =$

Try This! Write two multiplication and two division facts with the numbers 3, 4, and 12.

Let's Multiply and Divide

1. $8 \div 2 =$	2. $3 \times 2 =$	3. $15 \div 5 =$	4. $4 \div 2 =$	5. $0 \times 2 =$
6. $6 \div 2 =$	7. $9 \div 3 =$	8. $4 \times 4 =$	9. $4 \div 4 =$	10. $4 \times 3 =$
11. $10 \div 2 =$	12. $3 \times 1 =$	13. $12 \div 4 =$	14. $3 \times 4 =$	15. $2 \times 4 =$
16. $4 \div 1 =$	17. $5 \times 5 =$	18. $4 \times 2 =$	19. $20 \div 4 =$	20. $4 \div 2 =$
21. $3 \times 5 =$	22. $5 \div 1 =$	23. $16 \div 4 =$	24. $2 \times 3 =$	25. $8 \div 4 =$
26. $6 \div 2 =$	27. $12 \div 3 =$	28. $25 \div 5 =$	29. $10 \div 5 =$	30. $5 \times 2 =$
31. $12 \div 4 =$	32. $2 \times 2 =$	33. $20 \div 5 =$		
34. $3 \times 1 =$	35. $15 \div 3 =$	36. $5 \times 4 =$		

Slicing the Pie

Beside each pie write a fraction to show how much pie was eaten and how much was left.

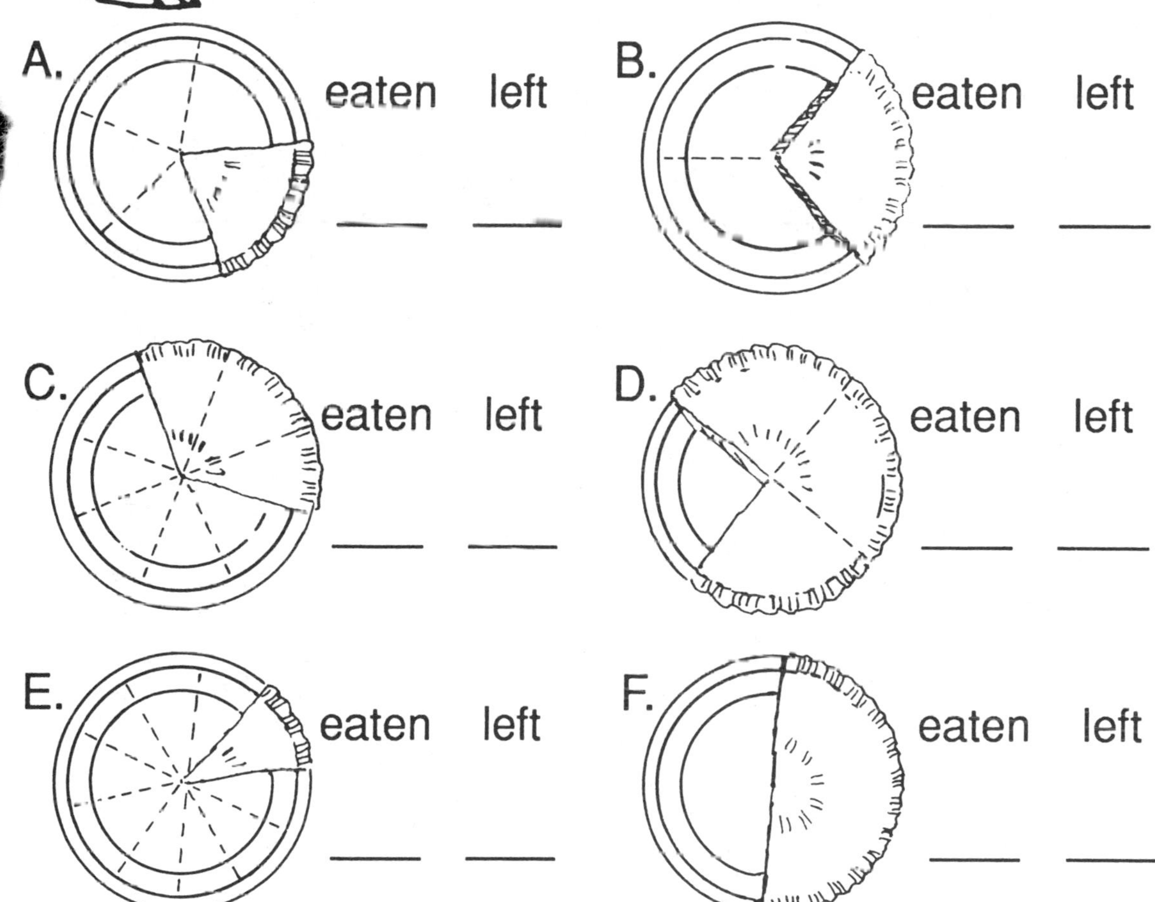

Try This! Draw two pies. Show one $\frac{3}{4}$ eaten and the other $\frac{6}{8}$ eaten. What do you notice?

More or Less?

Color to show each fraction. Then circle the fraction that is less.

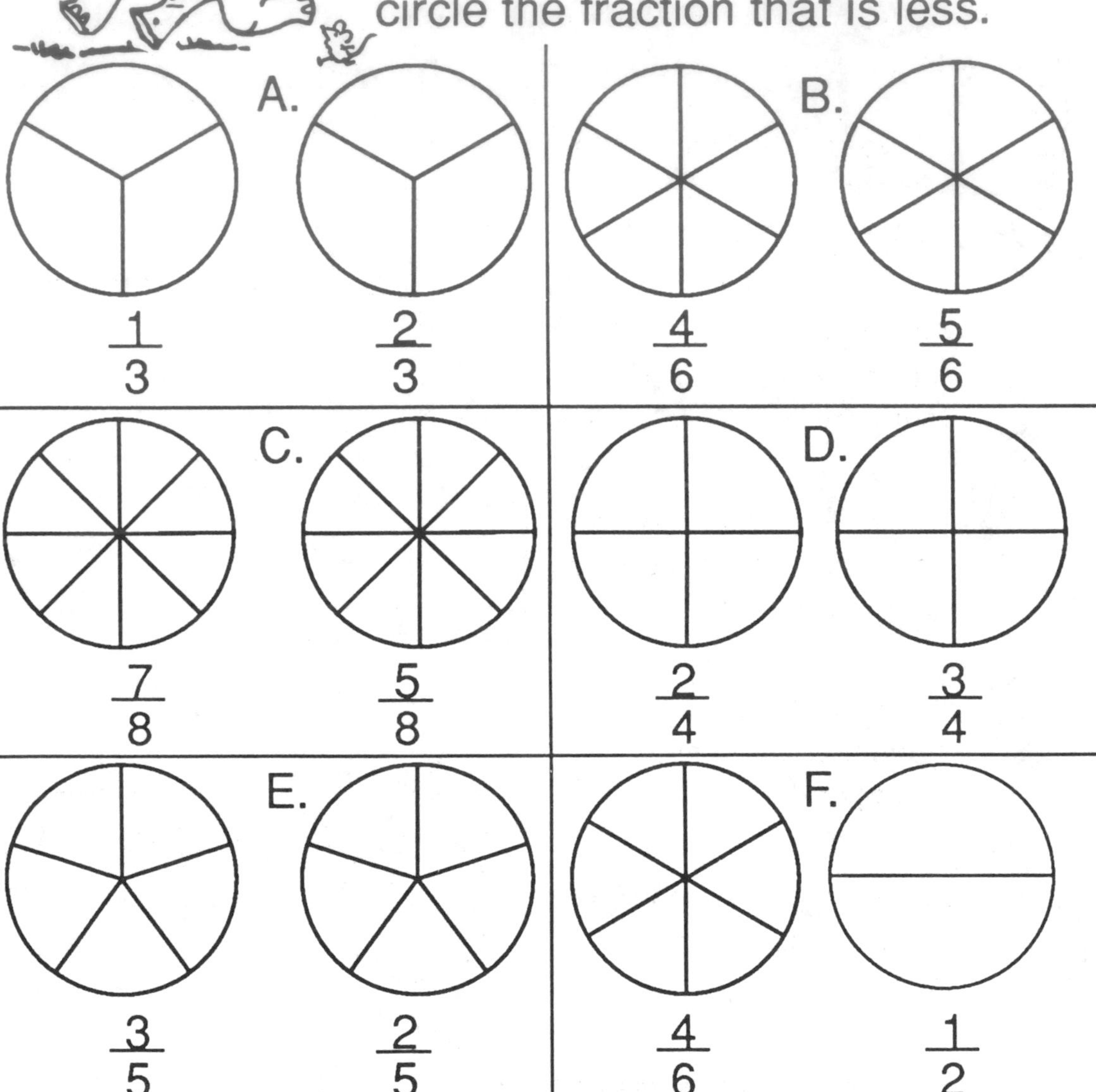

Try This! Write each pair of fractions above using a < or > sign. Circle the greater fraction.

Corners and Sides

Triangles have three sides and three corners.
A • shows each corner.
An **X** marks each side.

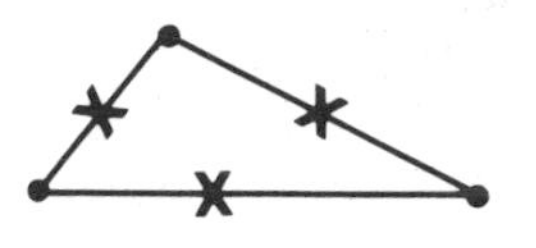

Write the letter of the shape that answers each riddle. Then label each shape with its name.

Shape Name

_____ 1. I have three straight sides and the same number of corners. I am a **triangle.**

_____ 2. I have four straight sides all the same length. I have four corners. I am a **square**.

_____ 3. I have four straight sides. I have four square corners. I am a **rectangle**.

_____ 4. I have five corners and five straight sides. I am a **pentagon**.

_____ 5. I have six corners and six straight sides. I am a **hexagon**.

_____ 6. I have no straight sides and no corners. I am a **circle**.

A

A. ____________________

B

B. ____________________

C

C. ____________________

D

D. ____________________

E

E. ____________________

F

F. ____________________

Try This! Draw a shape that has ten corners and ten sides.

Fun With Solid Figures

Find each of the shapes in the picture below.

cone

sphere

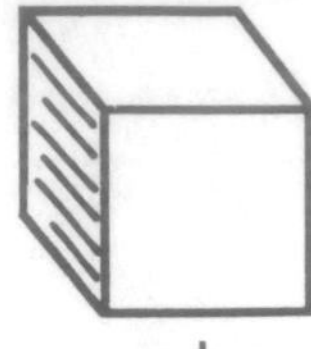
cube

cylinder

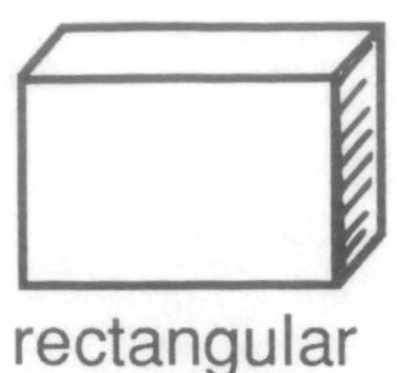
rectangular prism

Count the number of each shape in the picture. Record your answers.

1. ______ spheres
2. ______ cones
3. ______ cubes
4. ______ cylinders
5. ______ rectangular prisms

Color the picture using the code.

cones—red
spheres—blue
cubes—yellow
cylinders—orange
rectangular prisms—brown

Brainwork! Try drawing each of the solid figures above.

Homework Helper Record

Color the bean for each page you complete.

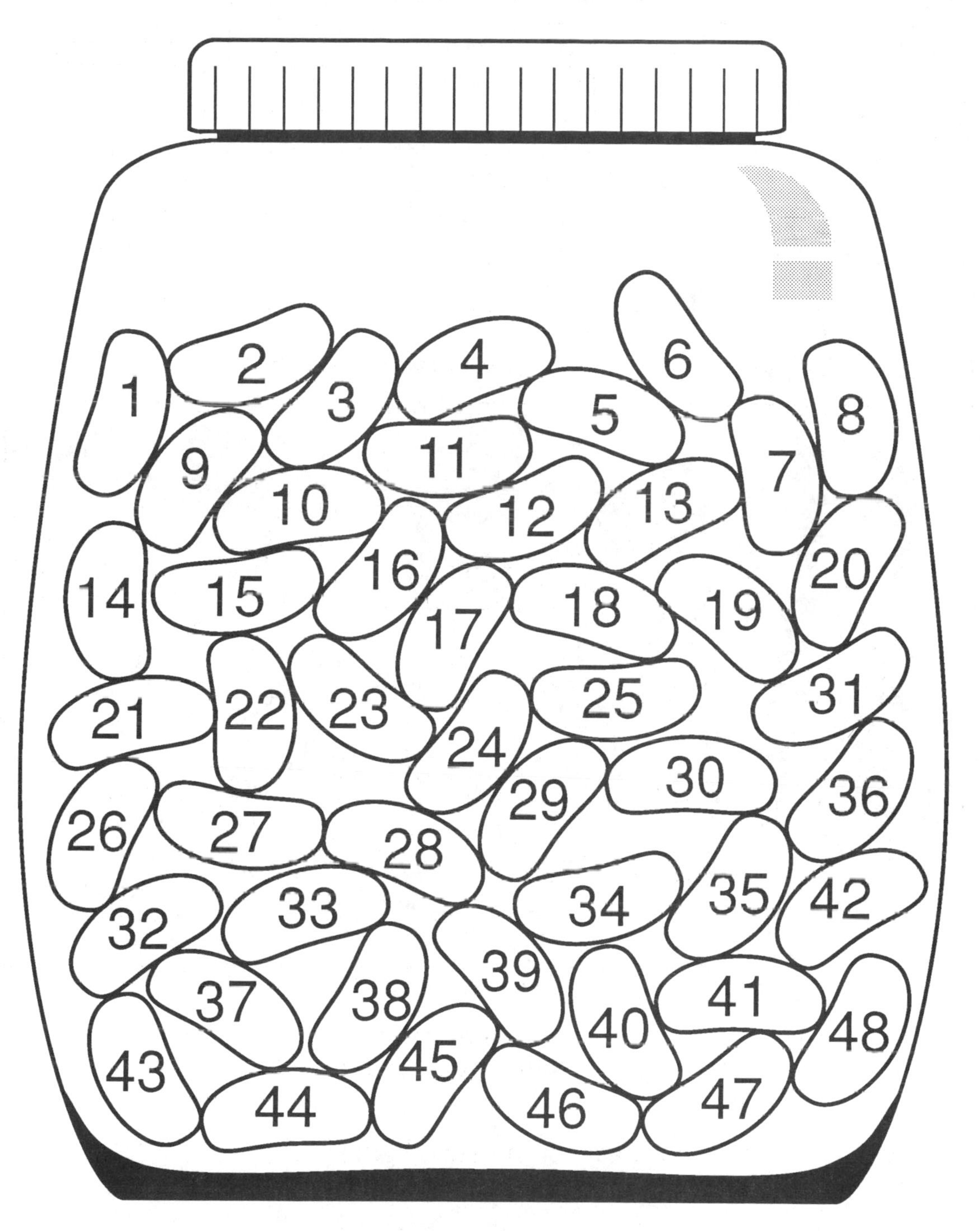

HOMEWORK AWARD
presented to
for successfully completing
this Homework Helper Book
signed
date